THE MICROARCHITECTURE OF PIPELINED
AND SUPERSCALAR COMPUTERS

T0191715

THE MICROARCHITECTURE OF PIPELINED AND SUPERSCALAR COMPUTERS

by

AMOS R. OMONDI

Department of Computer Science,
Flinders University,
Adelaide, Australia

KLUWER ACADEMIC PUBLISHERS

BOSTON / DORDRECHT / LONDON

A C.I.P. Catalogue record for this book is available from the Library of Congress.

ISBN 978-1-4419-5081-9

Published by Kluwer Academic Publishers,
P.O. Box 17, 3300 AA Dordrecht, The Netherlands.

Sold and distributed in North, Central and South America
by Kluwer Academic Publishers,
101 Philip Drive, Norwell, MA 02061, U.S.A.

In all other countries, sold and distributed
by Kluwer Academic Publishers,
P.O. Box 322, 3300 AH Dordrecht, The Netherlands.

Printed on acid-free paper

To my anchor:

Anne Hayes

and my mother:

Mary Omondi

Contents

Preface

This book is intended to serve as a textbook for a second course in the implementation (i.e. microarchitecture) of computer architectures. The subject matter covered is the collection of techniques that are used to achieve the highest performance in single-processor machines; these techniques center the exploitation of low-level parallelism (temporal and spatial) in the processing of machine instructions. The target audience consists students in the final year of an undergraduate program or in the first year of a postgraduate program in computer science, computer engineering, or electrical engineering; professional computer designers will also also find the book useful as an introduction to the topics covered. Typically, the author has used the material presented here as the basis of a full-semester undergraduate course or a half-semester postgraduate course, with the other half of the latter devoted to multiple-processor machines. The background assumed of the reader is a good first course in computer architecture and implementation — to the level in, say, *Computer Organization and Design*, by D. Patterson and H. Hennessy — and familiarity with digital-logic design.

The book consists of eight chapters: The first chapter is an introduction to all of the main ideas that the following chapters cover in detail: the topics covered are the main forms of pipelining used in high-performance uniprocessors, a taxonomy of the space of pipelined processors, and performance issues. It is also intended that this chapter should be readable as a brief "stand-alone" survey. The second chapter consists of a brief discussion of issues in timing and control: the topics covered are bounds on the processor clock-cycle, the implementation of clocking systems, the design of specialized latches for pipelining, and how to detect and deal with potential conflicts in resource usage. The third chapter deals with the implementation of high-performance memory systems and should largely be a review for well-prepared readers. The fourth, and longest, chapter, deals with the problem of ensuring that the processor is always adequately supplied with instructions; this is probably the most difficult problem in the design a pipelined machine. The fifth chapter deals with a similar, but more tractable, problem in the flow of data. The sixth chapter is an introduction to the design of pipelines to process vector data. And the last chapter covers mechanisms used to facilitate precise interruption.

Each chapter is divided into two main parts: one part covers the general principles and ideas, and the other part gives case studies taken from practical machines. A few remarks should be made regarding these case studies: First, they have all been selected to demonstrate particular points, although in a few cases there are similarities between the different machines covered; the reader therefore need not review at once all the case studies in any one chapter but may take different ones on different readings. Second, a requirement in their selection has been that they be adequately documented in the published literature; this has necessarily meant the exclusion of machines that would otherwise be of interest. Third, the reader should bear in mind that the latest machines do not necessarily employ fundamentally new techniques, nor is it the case that "old" machines necessarily employ "out-of-date" techniques: many of the sophisticated ideas in the implementation of current high-performance machines appeared twenty to thirty years ago — in machines such as the CDC 6600 and the IBM 360/91. Lastly, the amount of detail given on the various machines has been determined by what is available in the published literature, and this varies greatly; this has sometimes resulted in an uneven coverage, which should not be taken as an indicator of relative importance.

Acknowledgments

Gratitude to The Most Merciful, The Compassionate and Compassionating One, without whom no good thing is possible.

1 FUNDAMENTALS OF PIPELINING

We shall begin by introducing the main issues in the design and implementation of pipelined and superscalar computers, in which the exploitation of low-level parallelism constitute the main means for high performance. The first section of the chapter consists of a discussion of the basic principles underlying the design of such computers. The second section gives a taxonomy for the classification of pipelined machines and introduces a number of commonly used terms. The third and fourth section deal with the performance of pipelines: ideal performance and impediments to achieving this are examined. The fifth section consists of some examples of practical pipelines; these pipelines form the basis for detailed case studies in subsequent chapters. The last section is a summary.

1.1 INTRODUCTION

The basic technique used to obtain high performance in the design of pipelined machines is the same one used to obtain high productivity in factory assembly lines. In the latter situation, the work involved in the production of some object is partitioned among a group of workers arranged in a linear order such that each worker performs a particular task in the production process before passing the partially completed product down to the next worker in the line. All workers operate concurrently, and a completed product is available at the end of the line. This arrangement is an example of *temporal parallelism* — that

1

is, parallelism in time — and appears in several different forms in the designs of high-performance computers. The two basic forms of pipelining are *instruction pipelining* and *arithmetic pipelining*; other types of operations, e.g. memory accesses, can also be pipelined.

Figure 1.1. An instruction pipeline

Table 1.1. Flow in an instruction pipeline

Cycle No.	Stage 1	Stage 2	Stage 3	Stage 4	Stage 5
1	Instruction-1				
2	Instruction-2	Instruction-1			
3	Instruction-3	Instruction-2	Instruction-1		
4	Instruction-4	Instruction-3	Instruction-2	Instruction-1	
5	Instruction-5	Instruction-4	Instruction-3	Instruction-2	Instruction-1
6	Instruction-6	Instruction-5	Instruction-4	Instruction-3	Instruction-2
7	Instruction-7	Instruction-6	Instruction-5	Instruction-4	Instruction-3
8	Instruction-8	Instruction-7	Instruction-6	Instruction-5	Instruction-4
9	Instruction-9	Instruction-8	Instruction-7	Instruction-6	Instruction-5
⋮	⋮	⋮	⋮	⋮	⋮

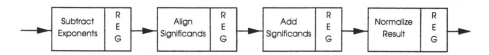

Figure 1.2. An arithmetic pipeline

An instruction pipeline consists of several hardware units, *stages* or *segments*, each of which is assigned some task in a sequence of tasks that collectively realize the complete processing of an instruction. For example, a simple partitioning of an instruction processing sequence might consist of five stages: FETCH INSTRUCTION, DECODE INSTRUCTION, FETCH OPERANDS, EXECUTE INSTRUCTION, STORE RESULTS. The corresponding pipeline is shown in Figure 1.1. (The pipeline consists of modules of combinational logic to compute the given subfunctions and registers to separate the inputs and outputs of different stages. We shall therefore view each stage as consisting of some subfunction logic and a register.[1]) To see the benefits of this arrangement, suppose each stage of the pipeline shown is capable of operating in 25 ns, that at the end of every 25 ns every instruction that is in the pipeline advances to the next stage, and that new instructions are fed into the pipeline at the same rate. Then the progress of instructions through the pipeline will be as shown in Table 1.1: Initially the pipeline is empty, and the first instruction that goes in must traverse the entire pipeline, taking 125 ns (25 ns × 5), before the processing of one instruction is complete and a result is available. But after this filling-up period, the pipeline is full, and one instruction is completed every 25 ns — every *cycle* (or *beat*). This is in contrast with an equivalent non-pipelined unit, in which an instruction is completed only every 125 ns at all times. The time required to completely process one instruction is the *latency* of the pipeline.

Arithmetic pipelines are less common than instruction pipelines. One gets such a pipeline by partitioning among the hardware stages the different phases of an arithmetic operation. As an example, Figure 1.2 shows a typical pipeline for floating-point addition; the processing of additions in this pipeline is similar to the processing of instructions in the pipeline of Figure 1.1. Arithmetic pipelines are typically used for floating-point operations in high-performance computers, although there are many instances where fixed-point operations too can be usefully pipelined. In the case of vector operations, pipelining both fixed-point and floating-point operations is especially advantageous, as the existence of a large number of operations of the same type can be guaranteed. Consequently, almost all vector-supercomputers employ arithmetic pipelining as the main technique for high performance.

1.2   A TAXONOMY OF PIPELINES

The pipeline shown in Figures 1.2 is capable of processing just one type of operation and is an example of a *unifunctional* pipeline. A *multifunctional* pipeline, on the other hand, is one that is capable of processing several types of operation. For example, each of main arithmetic pipelines in the Texas-Instruments Advanced Scientific Computer (Figure 1.3) is an example of a multifunctional pipeline: it can carry out addition, subtraction, multiplication, and division, all in both fixed-point and floating-point, as well as a variety of other more complex operations.

A pipeline is said to be *linear* if each stage is connected to just one other stage at its input and to one other stage at its output. A pipeline that is not linear is *planar*. The arithmetic pipeline in the Texas-Instruments Advanced Scientific Computer is planar, whereas both of the pipelines in Figures 1.1 and 1.2 are linear.

A pipeline may also be classified as being either *non-configurable*, if the connectivity of the stages is fixed, or *configurable*, if the connectivity is variable. In

---

[1] *Wave pipelining*, in which pipelines have no inter-stage registers, has recently become a promising area of research, although the basic idea has been around since the 1960s.

the latter case, each connection pattern essentially gives rise to a subpipeline of a distinct functionality. The two example pipelines of Figures 1.1 and 1.2 are non-configurable, whereas the arithmetic pipeline of the Texas-Instruments Advanced Scientific Computer is configurable. Within the category of configurable pipelines, we may further distinguish between *statically configurable* pipelines, in which only one active configuration exists at any given moment, and *dynamically configurable* pipelines, in which several active configurations can exist concurrently. The terms *static* and *dynamic* reflect the fact that in the one type of pipeline, configuration can take place only at the initiation of a job or very infrequently, whereas in the other, reconfiguration can take place on-the-fly and more frequently.

The instruction pipeline of Figure 1.1 can complete the execution of just one instruction in each cycle. It is, however, possible for a planar pipeline to have more than one unit in the EXECUTE stage and therefore be capable of completing the execution of several instructions in each cycle, provided these units can be supplied with instructions at a sufficiently high rate. A processor of this sort — that is, one that exploits *spatial parallelism* (i.e. parallelism in space) to initiate and execute several instructions in a single cycle — is a *superscalar* one [19, 20]. Superscalar processors are also commonly referred to as *instruction-level parallel* processors; although basic pipelining does involve some limited form of parallelism at the level of instructions, the term *instruction-level parallelism* is more commonly used to refer to just spatial parallelism. Many of the current high-performance microprocessors are superscalar; examples of these are given in Section 1.5. We shall say that a pipeline is k-*way superscalar* if it can complete the processing of up to k instructions in each cycle.

Consider again the pipeline of Figure 1.1. If this were to be modified to have just three stages — for example, FETCH-DECODE, FETCH-EXECUTE, and STORE — then (relative to the original pipeline) the cycle time would be greater than that required for a typical subfunction or (if the cycle time corresponds to a subfunction time) the number of instructions completed in each cycle would be less than 1 (relative to the rate of the original pipeline). In this case we would have an example of an *underpipelined* machine. The Stanford MIPS with three effective[2] stages (INSTRUCTION FETCH-INSTRUCTION DECODE; OPERAND DECODE/MEMORY STORE; MEMORY LOAD/READ REGISTERS, EXECUTE, WRITE REGISTERS) is an example of an underpipelined machine [14]. On the other, had the pipeline of Figure 1.1 been modified by further subdivision of each of the stages shown, and therefore had (relative to the original pipeline) a cycle time smaller than the time required for a typical subfunction, then we would have an example of a *superpipelined* machine [19, 20]. As an example, the 8-stage superpipeline of the MIPS R4000 (Section 1.5) is obtained by (essentially) replacing the one-stage instruction cache and one-stage data cache of the 5-stage MIPS R3000 pipeline with a two-stage instruc-

---

[2]Nominally the pipeline consists of six stages, but it can have no more than three active instructions at a time — in the even-numbered stages or in the odd-numbered stages.

tion cache and a three-stage data cache [25]. Other examples of superpipelined processors are the Manchester MU5 (Section 1.5), the Sun UltraSPARC (nine basic stages), and the Cyrix 6x86 (seven basic stages) [23, 26, 33]. And the DEC Alpha 21164 and 21264 are examples of machines that are both superscalar and superpipelined, as are several of the latest high-performance microprocessors [10, 12]. In most of these machines, the superpipelining is achieved through a finer subdivision of the memory-access, or arithmetic-execution stages, or both, although in a few cases it is possible to argue about whether or not they represent pure superpipelining.

So far we have implicitly assumed that once an instruction has gone through the pre-execution stages, it is immediately forwarded for execution. This, as we shall see in subsequent chapters, is frequently not the case, especially in superscalar pipelines. Accordingly, we shall sometimes distinguish between when an instruction is *dispatched*, which is when the pre-execution phases are complete, and when the instruction is *issued*, which is when it is forwarded for execution; in simple, linear pipelines, and those in which isntruations are dispatced with operands, usually dispatch and issue coincide.[3] Instructions may be issued in logical (i.e. program) order or *out-of-order*, dispatch is almost always in-order. Instructions can also *complete* execution out-of-order, depending on the availability at issue-time of execution units and the latencies of the operations involved. So, assuming in-order dispatch, the three possible issue-completion organizations are *in-order issue* with *in-order completion*, *in-order issue* with *out-of-order completion*, and *out-of-order issue* with *out-of-order completion* (Out-of-order processing is most commonly associated with superscalar pipelines.) Furthermore, as we shall see in Chapter 7, it may be necessary, for the handling of exceptions, to distinguish between the time that an instruction leaves an execution unit (i.e. when the basic encoded function has been carried out) and the time when its processing may be considered to be really ended. We shall say that an instruction is *retired*, and its results *committed*, when it has had its intended effect on the architectural (i.e. logical) machine-state.[4] The objective then, in exception-handling, is to always have, in effect, in-order retirement, even with out-of-order completion. A number of mechanisms have been devised to facilitate this, the most commonly implemented of which mechanisms is the *reorder buffer*.

There are several other dimensions in the design space of pipelined computers, but we shall consider just one more: the clocking strategy. The example given in Table 1.1 of the flow of orders[5] through a pipeline assumes that all stages in the pipeline perform their task within the same amount of time and that at the end of this period each order in the pipeline moves to the next stage. Such processing requires that all stages operate on clock cycles of the

---

[3]There appears to be no standard terminology, and some of the literature uses *dispatch* where we use *issue* or interchange the two.
[4]Some of the literature uses *completed* instead of *retired*.
[5]We shall use *order* as a generic term for any type of operation that is carried out in a pipeline, e.g. basic instruction processing, arithmetic execution, memory access, etc.

same duration, and this type of pipeline is therefore *synchronous*. In an *asynchronous* pipeline, on the other hand, the operational times of the stages may differ, with transfers between two stages taking place only when the one stage has completed its task and the other stage is ready to receive new input. An asynchronous pipeline may therefore require buffers to balance the flow between adjacent stages operating at different rates. Alternatively, an appropriate communication protocol (e.g. handshaking) may be used for each transfer between such stages. Unless otherwise stated, we shall henceforth assume that we are dealing with just synchronous pipelines.

1.3 IDEAL PERFORMANCE OF A PIPELINE

The short discussion above on pipeline performance implies that, under the conditions stated, the peak performance of a pipeline is inversely proportional to its cycle time. Accordingly, one measure of the performance of a pipeline is its *throughput*, which is the number of orders processed per unit of time; ideally, this is one order per cycle for a non-superscalar machine and k orders per cycle for a k-way superscalar machine. There is, however, an initial filling-up period before the pipeline attains its maximum throughput. This period, which is known as the *start-up* time, is proportional to the length of the pipeline, as is the *drain-time*, which is the time required to empty the pipeline after entry of the last order.

Another way to measure the performance of a pipeline is to calculate the improvement gained over a corresponding non-pipelined unit processing similar orders. A simplified analysis of this measure, which is known as the *speed-up* of the pipeline, is as follows. Suppose we have a non-superscalar pipeline of n stages operating on a cycle time of τ and that there are M orders to be processed. The first order into the pipeline will reach the other end of the pipeline after a time of $n\tau$, which is the start-up time. The remaining $M - 1$ orders will then be completed at the rate of one every τ and will therefore require a time of $(M - 1)\tau$. This gives a total time of $(n + M - 1)\tau$ for the M orders. Assuming that the pipeline was obtained from a straightforward partitioning of the non-pipelined version, we may take the clock cycle in the latter to be $n\tau$. The M orders will therefore require a time of $nM\tau$ if there is no pipelining. Hence the speed-up, S_M, for M orders is given by

$$S_M \triangleq \frac{nM\tau}{(n + M - 1)\tau}$$

and as more orders are processed

$$\lim_{M \to \infty} S_M = n$$

That is, in the limit, the pipeline of n stages will be n times as fast as the corresponding non-pipelined unit, and for a k-way superscalar machine with n stages, the speed-up may be taken as kn, relative to a corresponding non-pipelined, non-superscalar machine. A pipeline can, however, give even better performance than what we might expect from this sort of analysis, because the specialization of logic may allow some tasks to be performed faster: for example, all other things being equal, an addition carried out in a dedicated adder is usually faster than one carried out in a general-purpose arithmetic-and-logic unit. Nevertheless, for practical purposes, the best speed-up of a pipeline is useful only as an ideal to strive for: a real pipeline running real programs will usually have an *average speed-up* that is less than the maximum; similar remarks apply to throughput. (Both the average speed-up and the average throughput can be obtained from detailed simulations and benchmarking.) Furthermore, in most cases the time to process an individual order is likely to be increased by pipelining, as the pipeline registers typically introduce an overhead that is not in the equivalent non-pipelined unit; so the clock cycle might not decrease to the extent assumed above.

The quality of a pipeline may also be measured by calculating the utilization of the hardware. The *efficiency*, ρ_M, of an n-stage non-superscalar pipeline processing M orders is defined as

$$\rho_M \quad \stackrel{\triangle}{=} \quad \frac{speed-up}{number\ of\ stages}$$

$$= \quad \frac{nM\tau}{n(n + M - 1)\tau}$$

and as more orders are processed

$$\lim_{M \to \infty} \rho_M = 1$$

(For a k-way superscalar machine, we may use a similar definition, relative to k; that is, to divide ρ above by k.)

Performance analyses of the type given above, as well as the accompanying ideal figures, are of limited value, for the reasons already stated above and also because they do not easily transfer to the more complex designs of current high-performance machines. Average figures, calculated from empirical data are more useful, and in addition to those given above, one that has recently gained much currency is the *clock-cycles-per-instruction* (CPI), which is defined to be the average number of clock cycles required to execute an instruction in a typical program [13]. The CPI is therefore the inverse of the average throughput, and it too can be measured from detailed simulations and benchmarking. The best CPI for a non-superscalar pipeline is 1, and that for a superscalar pipeline is less than 1 and inversely proportional to the degree to which the machine is superscalar. A non-pipelined machine can also have a CPI of 1, but the main difference between a non-pipelined machine and a corresponding pipeline is that, even at the same CPI, the latter may have a faster clock. That is, the product *CPI×clock-cycle* (which is just the average time per instruction) will distinguish between the performances of the two and should be smaller for the pipelined version. Alternatively, we may say that given a non-pipelined machine with a CPI of λ, the ideal CPI for a corresponding n-stage pipeline should be λ/n (if the clock cycle is decreased proportionately). In current processor designs, the tendency is to try and achieve higher performance by increasing the effective degree to which a processor is superscalar, instead of by increasing the level of pipelining. This is because it is probably easier to reduce the CPI than to increase pipeline length and reduce cycle time, and a number of sophisticated techniques have been employed to the former end.

1.4   IMPEDIMENTS TO IDEAL PERFORMANCE

The simplistic analysis in the preceding section suggests that we can construct a pipeline of arbitrarily high performance simply by increasing the number of stages employed for the job at hand. There are several reasons why this is not so. First, pipeline length cannot be increased arbitrarily: there is a limit, and sometimes not a very high one at that, on the extent to which a function (instruction processing, arithmetic operation, etc.) can be partitioned into several subfunctions corresponding to the stages; moreover, the number of stages must always be integral. Second, the analysis above assumes that as pipeline length increases, the cycle time decreases proportionately. Such a change is not always possible — for example, because of the overheads introduced by inter-stage registers; bounds on the cycle time are discussed in Chapter 2. There are several other reasons why a pipeline might not consistently perform at its theoretical peak, and we shall discuss these briefly in the remainder of this section and in detail in subsequent chapters.

For performance measures such as speed-up, throughput, and CPI, it is not unusual for the the *averages* to be well below the theoretical maximums. The reasons for this will be evident from an examination of the assumptions inherent in the analysis above. One of these assumptions is that a new order is fed into the pipeline at every cycle. This assumption does not always hold. In an instruction pipeline, the assumption may not be valid whenever a branch[6] takes place and new instructions must be fetched from a memory whose access time is (typically) much greater than the pipeline's cycle time; the start-up time then shows up again. And in an arithmetic pipeline, the assumption can fail to hold because some of the orders in a typical program are control instructions, the processing of which will disrupt the flow. The problem of keeping a pipeline adequately supplied with instructions is discussed in detail in Chapter 4. One of the major techniques discussed in that chapter is that of *branch prediction*, which involves making an early guess of the target of a branch instruction. The prediction will be incorrect in some cases, and some instructions may be processsed (up to completion but before retirement) before it is ascertained that they should or should not actually be processed; this is appropriately known as *speculative processing* and generally requires that branch mispredictions be handled in the same manner as interrupts and other exceptions.

---

[6]Unless otherwise specified, we shall use *branch* to refer to any type of transfer of control.

Another assumption made in the above analysis of performance is that in every cycle, every order that is in the pipeline progresses from one stage to the next. This movement is possible only if all the orders in the pipeline are independent, and a cursory examination of a typical program will reveal that this is almost never the case. For example — see Table 1.2(a) — in an instruction pipeline, if instruction $i + 1$ uses as an operand a value that is computed by instruction i, then there is a stage (say, the operand-fetch stage) beyond which the former instruction cannot proceed until the result is available from the latter instruction; consequently, a bubble of empty stages appears in the pipeline and the throughput drops. This type of conflict, of which Table 1.2(a) shows three different types, is known as a *data dependence* or *data hazard*. Many, but not all, such hazards can be eliminated by employing the technique of *renaming*, in which the operand source and destinations names supplied by the programmer or compiler are replaced by other names made available at compile-time or (more usually) at run-time. An example of (static) renaming is given in Table 1.2(b), in which the renaming has eliminated the original hazards between the first and third instructions, between the second and third instructions, and between the first and fourth instructions. In dynamic (i.e. hardware-implemented) renaming, usually a distinction is made between the the *architectural-register* names (i.e. those specified in the original code) and the *implementation-register* names (i.e. those that refer to the actual storage), and the renaming then consists of mapping the architectural names to the implementation names. In such a case the number of physical registers need not be equal to the number of architectural names. Dynamic renaming can also be applied to memory addresses, but this is not as common as register renaming.

Table 1.2.   Renaming to eliminate hazards

Original code	Renamed code
R0 ⇐ R1/R2	R0 ⇐ R1/R2
R3 ⇐ R0*R4	R3 ⇐ R0*R4
R0 ⇐ R2+R5	R7 ⇐ R2+R5
R2 ⇐ R0–R3	R8 ⇐ R7–R3
(a)	(b)

Evidently, renaming is effective only with hazards that arise from (artificially) shared result-destination operand-names. For hazards that arise when one instruction depends on another for an operand, there is little that can be done beyond minimizing any waiting time, as well as ensuring that stalled instructions do not affect the flow of other instructions in the pipeline. Thus *data forwarding* consists of sending data directly to a waiting instruction and so eliminating the time that would otherwise be spent in reading from intermediate storage. Another commonly implemented technique (which employs forwarding) is to shunt waiting instructions into temporary storage known as *reservation stations*, thus allowing following, independent, instructions to proceed and to then send data (when available) directly to these stations. The instructions are then issued out-of-order from the reservation stations, according to the availability of operands and execution units. An alternative to reservation stations is to use simple buffers that hold only instructions (i.e. no data) but which still permit out-of-order issue; the drawback of these is that operands must then still be read from the register file before execution.

An order can also fail to advance to another stage if that stage is already in use or if the resources needed by the stage are already in use. For example, in an instruction pipeline, if the COMPUTE-EFFECTIVE-ADDRESS stage and the EXECUTE stage share an adder, then there will be instances where an instruction at one of the two stages has to wait for an instruction at the other stage to release the adder; the same thing might happen in, for example, a shared arithmetic pipeline in which a floating-point ADD and a floating-point MULTI-PLY share an adder and concurrently require access to it. (The Stanford MIPS, cited above as an example of an underpipelined machine, has six actual stages but only three effective stages because of such conflicts.) Another example of a similar resource conflict is the case of concurrent access, by different stages, to a register file with an insufficient number of read/write ports. A conflict of the type given in these three examples is known as a *collision*, or a *structural dependence*, or a *structural hazard*. Such a hazard can be easily avoided, or its likelihood reduced, by replicating resources — for example, by increasing the numbers of adders or read/write ports in the examples just given. We shall study structural hazards in more detail in Chapter 2, and data hazards constitute the main subject of Chapters 5.

1.5 CASE STUDIES

We now give several examples of practical pipelines.[7] The first two examples are of arithmetic pipelines; these are the VAX 6000/400 arithmetic pipeline and one of the four arithmetic pipelines in the Texas Instruments Advanced Scientific Computer (TI-ASC). The other pipelines covered incorporate both instruction pipelining and arithmetic pipelining and are superpipelined, or superscalar, or both. We shall use the second group of examples as the basis of several detailed studies throughout the text as we discuss various aspects of pipelined machines in more detail. The machines are the Manchester MU5 (from which the machines at the high-end of the performance range in the ICL 2900 series are derived), the DEC Alpha 21164 and 21264 (two implementations of the DEC Alpha AXP architecture), the PowerPC 604, Power PC620, and Power3 (implementations of the PowerPC architecture), the MIPS R4000 and

---

[7]For introductory purposes, a general, high-level understanding of these will suffice, and not all of the case studies need be reviewed at once. Details are discussed in later chapters, and the reader may wish to refer to this section only as necessary.

MIPS R10000 (implementations of the MIPS architecture), the Advanced Micro Devices AMD K6 (an implementation of the x86 architecture), and the Intel Pentium II and Pentium Pro (also implementations of the x86 architecture).

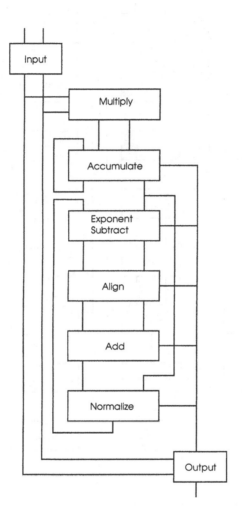

Figure 1.3.   TI-ASC arithmetic pipeline

1.5.1 TI-ASC arithmetic pipeline

The TI-ASC arithmetic pipeline is planar, multifunctional, and configurable, with the organization shown in Figure 1.3 [6, 35]. The pipeline has eight main stages:[8] (1) The INPUT stage is used to buffer data between the memory units and the arithmetic unit and to route arithmetic results directly back into the pipeline.

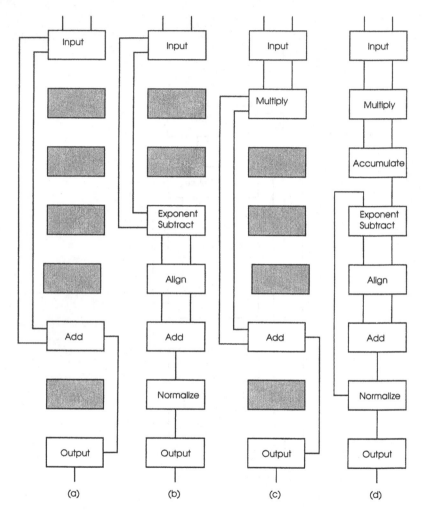

Figure 1.4. Configurations of the TI-ASC arithmetic pipeline

(2) The MULTIPLIER stage is used to multiply fixed-point numbers and significands of floating-point numbers and also to carry out division (by multiplicative normalization). The multiplier is a Wallace-tree with multiplier-recoding, and its final outputs are partial-sum and partial-carry vectors. (3) The ACCUMULATE stage consists of a 64-bit carry-lookahead adder that adds two or three

---

[8]The reader who is not familiar with the design of arithmetic units may find it useful to refer to [27].

numbers at a time. The stage is also used to assimilate the outputs of the MULTIPLY stage. (4) The EXPONENT-SUBTRACT stage is used to determine which of two floating-point operands has the smaller exponent and what the exponent-difference is. This stage also contains logic for all fixed-point and floating-point comparison orders. (5) The ALIGN stage carries out the alignment shifting for floating-point addition and subtraction and also performs all other operations that require a right-shift. (6) The ADD stage consists of a 64-bit carry-lookahead adder that is used for all primary additions and subtractions in both fixed-point and floating-point operations. (7) The NORMALIZE stage is used for normalization in floating-point operations and also for all operations that require a left-shift. (8) The OUTPUT stage is used to route results to their destination and to carry out logical operations.

Figure 1.4 shows four different configurations of the TI-ASC pipeline: for (a) fixed-point addition, (b) floating-point addition, (c) fixed-point multiplication, and (d) vector inner-product.

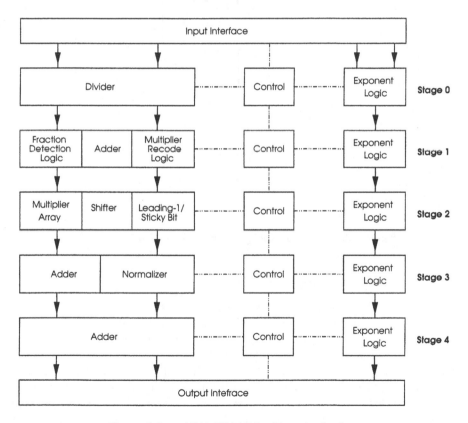

Figure 1.5.   VAX 6000/400 arithmetic pipeline

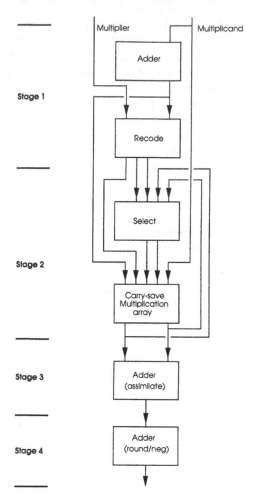

Figure 1.6.   Multiply-configuration of VAX 6000/400 pipeline

1.5.2 VAX 6000/400 arithmetic pipeline

The arithmetic pipeline of the VAX 6000/400 is another example of a configurable, multifunctional pipeline [9]. The unit has the organization shown in Figure 1.5 and is used to carry out floating-point addition, subtraction, division, and multiplication. The significand datapath is on the left in the figure and is sixty bits wide; the exponent datapath is on the right and is thirteen bits wide. (The description that follows is limited to the significand datapath.)

The first stage of the pipeline is used by divide operations only. The stage consists of an array-divider, plus associated quotient logic, and implements a high-radix SRT algorithm. The second stage is used for 60-bit addition and subtraction, to form multiplicand-multiples for multiplication, and to assimilate partial-carry and partial-sum vectors during division. The third stage consists of a radix-8 recoding Wallace-tree multiplier, a 57-bit shifter for significand alignment, a leading-1s detector, and logic to form sticky bits for rounding. The fourth stage consists of a 57-bit shifter for normalization and a 60-bit adder used for the addition and subtraction of significands and for assimilating partial-sum and partial-carry vectors during multiplication. The last stage consists of a 60-bit adder used for rounding and negation. The pipeline configuration for multiplication is shown in Figure 1.6, and that for addition is shown in Figure 1.7.

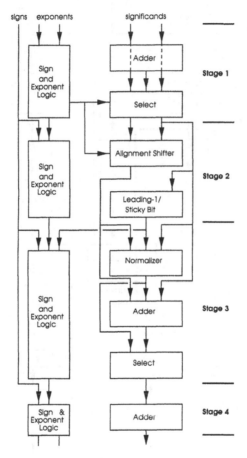

Figure 1.7.   Add-configuration of VAX 6000/400 pipeline

1.5.3 MU5 instruction pipeline

The MU5 processor pipeline has two main components: the Primary Instruction Pipeline, which carries out the preprocessing of all instructions, and the Secondary Instruction Pipeline, which completes the processing [26].

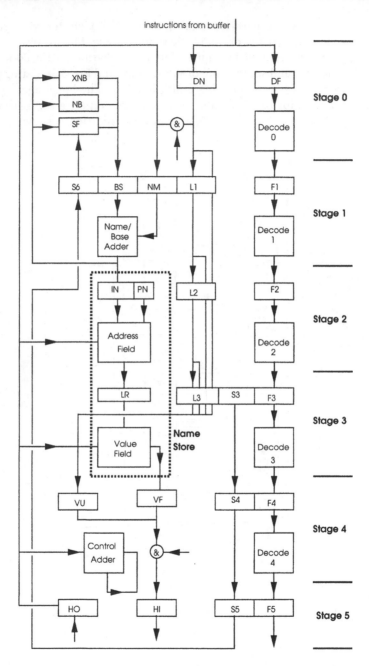

Figure 1.8.   MU5 Primary Operand Unit

The two pipeline structure was developed to match operand-access in typical high-level languages: the first pipeline is for accessing named operands (scalars and descriptors of arrays, strings, etc.), and the other pipeline is for accessing elements within data structures. The main unit in the Primary Instruction Pipeline is the Primary Operand Unit (PROP), whose organization is shown in Figure 1.8. PROP receives instructions from the Instruction Buffer Unit (which is described in detail in Chapter 4) and carries out the initial decoding and operand access. Instructions leaving PROP are either ready for execution or require another operand access (in the Secondary Instruction Pipeline) before they are.

An instruction entering PROP is partitioned into two parts: the opcode (into the register DF) and the operand-specifier (into the register DN); the latter specifies a displacement from some base. The first stage of the pipeline is nominally the Decode stage; however, only the initial decoding is carried out at that stage, and decoding logic is spread throughout the pipeline as needed. The first stage also includes the selection of one of three base registers (XNB, NB, SF): an operand may be global, local to a subroutine, or a parameter on the stack. In the second stage, the contents of DN are added (possibly after a shift) to the contents of the selected base register and the result used to access the Name Store (an operand buffer) in the third and fourth stages. (The Name Store is described in more detail in Chapter 3.) The fifth stage is the Operand-Assembly stage, which properly positions operands for input to the data highway. The sixth stage is the Highway-Input stage. Once an instruction leaves PROP, the Control Adder is used to increment the program counter.

The other units shown in Figure 1.8 are a variety of registers, most of which serve the usual role of inter-stage holding posts. The exceptions are the registers L1, L2, L3, S3, S4, S5, and S6: the L registers are used in the assembly of long literals, and the S registers are used to hold old values of the Stack-Front (SF) register should a rollback be required if incorrect instructions follow a branch instruction into the pipeline and their execution modify the contents of that register.

1.5.4 DEC Alpha 21164 and Alpha 21264

The DEC Alpha 21164 microprocessor is a superscalar second-generation (following on the Alpha 21064) implementation of the DEC Alpha AXP architecture, and the Alpha 21264 is the next implementation [10, 12, 29]. The Alpha 21164 pipeline nominally consists of five main stages — FETCH INSTRUCTIONS, DECODE, FUNCTIONAL-UNIT SELECTION, ISSUE/READ OPERANDS, EXECUTE/WRITE-BACK — but the EXECUTE stage is further pipelined, so that the effective pipeline length is seven for fixed-point arithmetic instructions, nine for floating-point arithmetic instructions (except for division), and twelve for on-chip memory operations. The complete organization of the processor is shown in Figure 1.9, and Figure 1.10 gives the pipeline flow for the different types of instructions. The main components of the processor are the Instruction Unit, the Integer Execution Unit, the Floating-Point Unit, the Memory Unit, and the Cache-Control and Bus-Interface Unit.

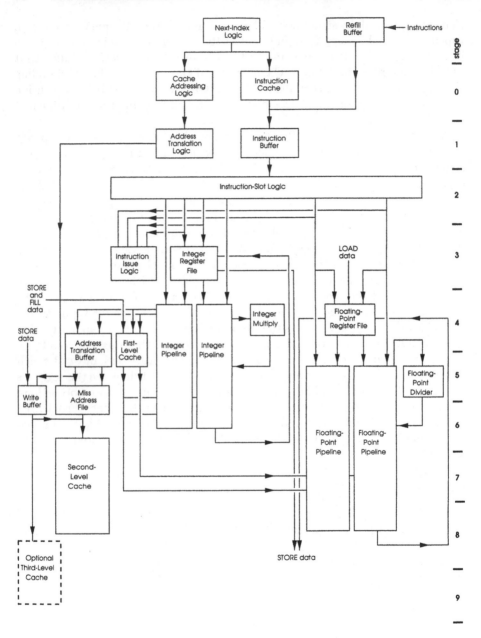

Figure 1.9.   DEC Alpha 21164 processor pipeline

The first four stages of the pipeline above make up the Instruction Unit, whose task is to fetch, decode, and issue (in program order) instructions to the execution units. The first stage consists of an instruction cache and associated addressing logic. The second stage consists of an instruction buffer and virtual-address translation buffers. The third stage consists of next-instruction-address logic and execution-unit-selection logic. And the fourth stage consists of instruction-issue logic and registers for fixed-point computations. The operation of the Instruction Unit is discussed in more detail in Chapter 4.

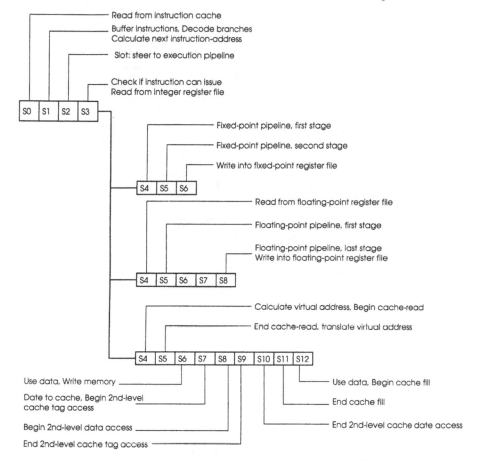

Figure 1.10.   DEC Alpha 21164 pipeline flow

The processor has six arithmetic execution units. The fixed-point units are a multiplier (nominally located in the fifth stage of the pipeline but used in conjunction with logic in the rest of the arithmetic pipeline) and two multi-purpose units (spread through the fifth, sixth, and seventh stages of the pipeline). The floating-point units are a divider (nominally located in the sixth stage of the pipeline but consisting of logic that is used iteratively together with logic in the rest of the arithmetic pipeline) and an add/divide unit and multiplier, both spread through the sixth through the ninth stages of the pipeline. (The fixed-point and floating-point register files are located in the fourth and fifth stages of the pipeline.) Two levels of cache are used for data: the first-level cache (and associated buffers) is in the fifth through seventh stages of the pipeline, and the second-level cache is in the seventh through ninth stages. The ninth stage also includes the cache control and interface logic.

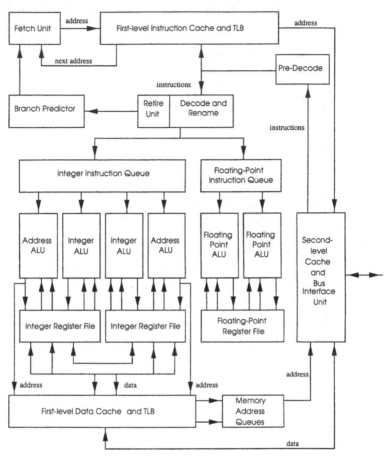

Figure 1.11.   DEC Alpha 21264 processor pipeline

The DEC Alpha 21264 is a performance-enhancement extension of the Alpha 21164: several changes have been made in both the implementation and the realization. The major implementation features introduced in the Alpha 21264 are two more functional units (for "address" computations), register renaming, an increase in the size of the "window" of concurrently active instructions (up to 80), with the ability to issue more instructions in each cycle — up to six at peak and up to four sustainable — branch prediction, and out-of-order instruction-issue. The caches in the new processor are also much larger and are fully pipelined.

The general organization of the Alpha 21264 is shown in Figure 1.11. (Functionally, the pipeline layout is similar to that of the Alpha 21164.) Instructions are initially entered in the instruction cache, after being pre-decoded to determine the type of functional unit required in the execution phase; the type-information is appended as three extra bits to each instruction. Instructions are retrieved from the instruction cache, decoded and renamed (as necessary, in a system that includes 41 extra integer registers and 41 extra floating-point registers), and then dispatched to one of two "queues", according to type. From the queues, instructions are issued (out-of-order) to appropriate functional units, with operands read from and results returned to the corresponding register file. (The integer register file has been replicated because a single register file with eight read ports and six write ports could not be realized without compromising performance.) The execution units consist of two general-purpose integer units that carry out arithmetic and logical operations, shifts and branches, and motion-video operations; two address ALUs that carry process LOADs, STOREs, and simple arithmetic and logical operations; and two floating-point units. Lastly, the Retire Unit retires instructions in logical order if they complete execution without causing an exception.

1.5.5 PowerPC 604, PowerPC 620, and Power3

The PowerPC 604 is one of the most sophisticated implementations of the PowerPC architecture; it incorporates many of the most advanced techniques used in the design of pipelined superscalar processors [31]. The organization of the processor is of a pipeline with six main stages, as shown in Figure 1.12.

The Fetch stage translates instruction-fetch addresses and reads instructions (four at a time) from the instruction cache into the pipeline; the stage also contains a branch-prediction unit. The Decode stage carries out the instruction decoding and determines what resources are required by an instruction. The Dispatch stage determines what execution units are required and forwards the instructions, at a rate of up to four in each cycle, to reservation stations associated with the execution units. The Dispatch stage also contains additional branch-prediction logic whose results are more accurate than those from the Fetch stage: the Fetch-stage prediction is used to fetch instructions early enough but may later be corrected by the Dispatch-stage prediction. Since the base processing rate of the Fetch unit exceeds that of the Dispatch Unit, the 8-entry Instruction Buffer is employed to balance the flow between the two units. The Execution stage consists of six independent functional units: three fixed-point arithmetic units, one floating-point arithmetic unit, one Load/Store unit (for accesses to memory), and a unit to execute branch instructions; all of these units are pipelined. Each functional unit has associated reservation stations to hold instructions that are waiting for the unit to become free or for operands to become available, and instructions may be issued and executed out-of-order. The out-of-order processing of instructions means that special action is required to ensure that exceptions can be handled precisely. The Retire stage (by means of a 16-entry buffer) keeps track of the original order of the instructions, as well as their status, as they go through the pipeline and, once execution is completed, forwards them to the Writeback stage, where they are retired in program-order. The registers associated with the Writeback stage consist of implementation registers that correspond to the architectural registers and "shadow registers" that are used during renaming; the assignment of the shadow registers is made during the Dispatch stage. Results coming from the execution units are held in the shadow registers until it is safe to commit them into the final destination registers. The Retire stage also detects and processes exceptions.

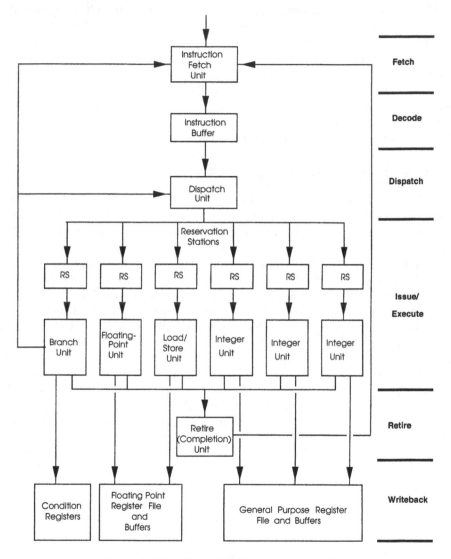

Figure 1.12.   PowerPC 604 processor pipeline

The PowerPC 620 is a following implementation of the PowerPC architecture [8]. The main differences between this implementation and the PowerPC 604 are that the former is a 64-bit implementation, has two large on-chip caches, has an improved bus-interface and memory system, and uses more aggressive techniques to deal with branch instructions. Other, lesser, differences are noted in later chapters.

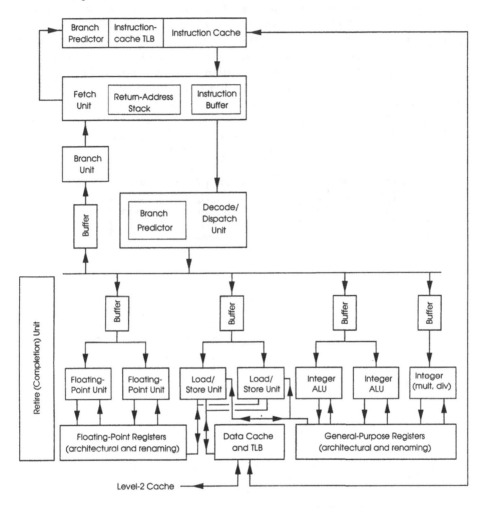

Figure 1.13.   Power3 processor pipeline

The PowerPC 620 has now been superseded by the Power3, which is, essentially, a derivation of the former microarchitecture [32]. The new aspects of the latter include two more execution pipelines (a floating-point unit and a load-store unit), more buffers and renaming registers, performance-enhancement changes to the cache-memory system, and a replacement of the sophisticated reservation stations with simpler buffers that hold instructions but not data. The general organization of the Power3 is shown in Figure 1.13. In each cycle up to eight instructions are fetched from any position within a 128-byte cache line and placed into the 16-entry Instruction Buffer. Branch prediction is used, as necessary, within this stage, which also includes an 8-entry Return-Address

Stack for subroutines. (As with its predecessors, the Power3 employs additional branch prediction in the Decode/Dispatch stage.) The Decode/Dispatch stage then processes (with renaming) up to four instructions in each cycle and dispatches these, in logical program order, to buffers associated with the execution units. Instructions are issued out-of-order from these buffers, and, with eight execution units, the execution of up to eight instructions can be started in each cycle. Instructions completing execution store their results in temporary registers allocated during renaming. The Retire Unit then retires instructions, up to four per cycle, by committing results from the temporary registers into the registers that define the architectural state. (The registers are realized as a replicated set, in order to minimize structural hazards without increasing the number of read/write ports on a single register file.)

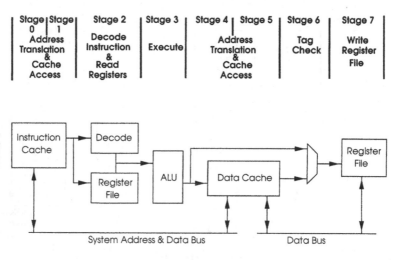

Stage 0	Stage 1	Stage 2	Stage 3	Stage 4	Stage 5	Stage 6	Stage 7
Address Translation & Cache Access		Decode Instruction & Read Registers	Execute	Address Translation & Cache Access		Tag Check	Write Register File

Figure 1.14.   MIPS R4000 processor pipeline

1.5.6 MIPS R4000 and MIPS R10000

The MIPS R4000 (an implementation of the MIPS architecture) is a simple example of a superpipelined processor [25]. The basic organization of the pipeline is shown in Figure 1.14. The processing of an instruction starts with an access to the cache in the first two stages of the pipeline. The cache requires two cycles for a complete access, but it is pipelined (into one cycle for addressing functions and one cycle for the actual reading) and so can initiate a new access in every cycle.[9] In the third stage, the instruction is decoded and source operands are read from the register file. Arithmetic operations are then ready for execution in the fourth stage. The fifth, sixth, and seventh stages are for accessing the

---

[9]Strictly, part of the addressing functions (the tag checking) is overlapped with the operations in the third stage.

data cache, which is managed in a manner similar to the instruction cache. The results of instruction execution are written back to the register file in the eighth stage.

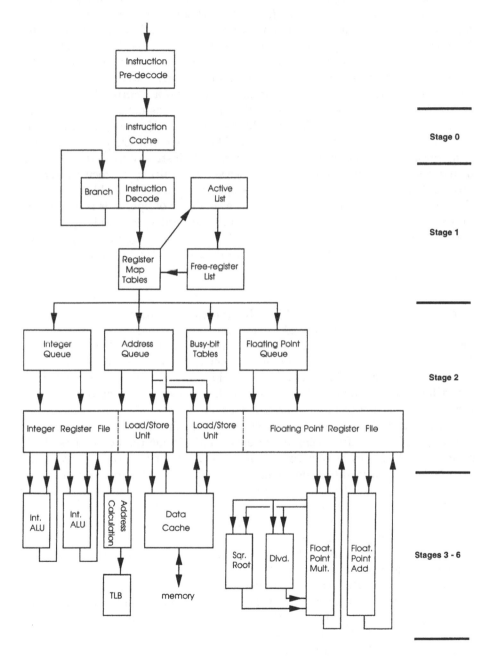

Figure 1.15.   MIPS R10000 processor pipeline

The MIPS R10000 is a more recent implementation of the MIPS architecture [24, 37]. It is a pipelined superscalar machine capable of issuing and executing up to four instructions in each cycle. The high-level organization of the machine is shown in Figure 1.15. The first three stages of the pipeline are the instruction-fetch stages. In the first stage, four instructions per cycle are fetched from a line of the instruction cache and then aligned. The logic in the second stage decodes the instructions and also includes logic for hazard-detection and renaming to eliminate hazards: the Register Map Tables associate architectural-register names with implementation-register names, the Active List identifies the implementation registers that are currently in use, and the Free-Register list identifies the implementation registers that are available for allocation in renaming. Target-address calculation for branch instruction also takes place at the second stage. The third stage consists of register files and "queues" (actually non-FIFO buffers) to hold instructions that have been dispatched (in-order) from the second stage and are waiting to execute but cannot yet do so, either because their source operands are not available or because the execution units required are busy; the availability of operands is monitored in a separate table. Instructions are issued, out-of-order, from this stage once operands and functional units are available. The remaining four stages make up the execution stages, with the cache also located in the last stage.

1.5.7 AMD K6

The Advanced Micro Devices AMD K6 (an implementation of the x86 architecture) is another recent high-performance superscalar processor [1]. The processor consists of a six-stage pipeline, in the organization shown in Figure 1.16, and can issue up to six instructions per cycle, with out-of-order execution. Renaming is used for data hazards, and a sophisticated branch-prediction algorithm has been implemented to reduce the negative effects on performance that branch instructions might otherwise have.

Bytes of code from memory enter the processor through the (first-level) instruction cache just after being predecoded to determine the boundaries of instructions, which may be of variable length; the boundary information from the decoding is stored in the cache, along with the instructions. Instructions may also be stored in the Branch Logic unit (which contains the branch-prediction logic), if they are the target of a branch instruction, and would then be fetched from there instead of from the main memory. The next steps in the flow of instructions take them through the 16-entry Instruction Buffer and into the Decode stage, where three decoders translate the variable-length "CISC" instructions into fixed-length "RISC" instructions that are then stored in the 24-entry Scheduler Buffer; only one decoder operates in each cycle, in which up to four "RISC" instructions are generated. Instructions are issued from the Scheduler Buffer under the direction of the Instruction Control Unit, which consists of logic to detect hazards and to rename operands, as appropriate; the unit can issue up to six instructions in each cycle and retire up to four in the same period. The execution of instructions may be out-of-order and takes place in eight functional units: a Load Unit (for memory-read operations), a Store Unit (for memory-write operations), two Integer units, a Floating-Point unit, a Branch Unit (for the resolution of conditional branches), a Multimedia Extension (MMX) Unit (for instructions used to support multimedia operations), and a 3DNow unit (for graphics). The processor's register file consists of 24 general-purpose registers that correspond to architectural general-purpose registers, 24 registers for general-purpose-register renaming, 9 registers that correspond to architectural multimedia registers, and 9 registers for multimedia-register renaming.

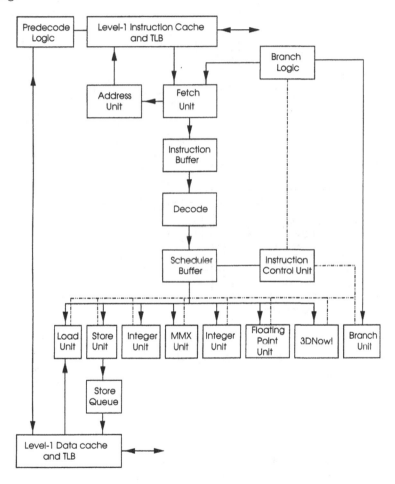

Figure 1.16.   AMD K6 processor pipeline

1.5.8 Intel Pentium Pro and Pentium II

The Pentium Pro is one of the latest implementations of Intel's x86 architecture, and it has some broad similarities with the AMD K6 implementation, which also implements the same architecture [18]. The processor is superscalar and has the general organization shown in Figure 1.17. Instructions enter the pipeline via the Instruction Fetch Unit, which fetches one 32-byte line per cycle from the Instruction Cache, marks the beginning and end of each instruction, and then transmits a sequence of aligned bytes to the Instruction Decoder; the fetch-phase involves the use of the Next-IP unit (which determines the address of the next instructions to be fetched) and branch prediction on the target addresses of branch instructions.

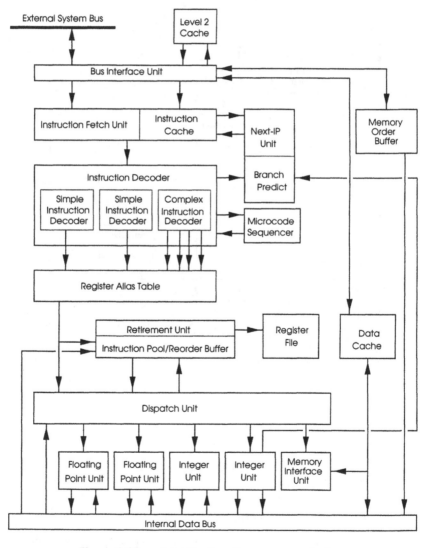

Figure 1.17.   Intel Pentium Pro processor pipeline

The Instruction Decoder consists of three parallel decoders — two for simple instructions, and one for complex instruction — that convert each incoming variable-length "CISC" instruction into a sequence of fixed-length "RISC" microinstructions, each of two source operands and one destination operand. In each cycle, each of the simple-instruction decoders can generate one microinstruction, and the complex-instruction decoder can generate up to four microinstructions. The microinstructions from the decoder are forwarded to the Register Alias Table (essentially a register-renaming and hazard-detection unit), where status and various flag bits are appended to them in order to facilitate out-of-order execution. The microinstructions are then dispatched into the Instruction Pool/Reorder Buffer, which contains microinstructions that are either waiting to be executed or that have been executed but have not yet had their results committed to the register file. The pool/buffer essentially combines the functions of reservation stations with those of a mechanism used to ensure precise interruptions.

Microinstructions enter the pool/buffer in logical program order but, under the direction of the Issue Unit, leave it (for execution) out-of-order, as permitted by data hazards and the availability of execution units. The Issue Unit continually scans the pool/buffer for instructions that are ready for execution and sends these, at a rate of up to five in each cycle, to execution units. Of the execution units, one of the two Integer units also processes branch microinstructions; in the event that a misprediction is made of the target of a branch, and an incorrect sequence of instructions fetched for processing, that Integer unit changes the status of corresponding microinstructions so that they are effectively removed from the pool/buffer. The Memory Interface Unit processes all LOADs and STOREs to and from the main memory and has two ports that enable it to process both a LOAD and a STORE in the same cycle; the associated Memory Order Buffer, which functions as an Issue Unit and Reorder Buffer for LOAD and STORE instructions to the second-level cache and main memory, aims to achieve, by reordering instructions, an optimal sequencing of memory operations while still ensuring consistency. Lastly, the Retirement Unit commits results to the register file and retires (in logical program order) the corresponding instructions; the unit does this by continually scanning the pool/buffer for microinstructions that have been executed without exceptions (including branch mispredictions) and that no longer have dependences with any other instructions in the pool/buffer. Up to three microinstructions are retired in each cycle.

The Pentium II has a microarchitecture that is largely similar to that of the Pentium Pro but in a different (and faster) realization and with a few microarchitectural changes that include larger instruction and data caches and more execution units. The additional units are two Multimedia Extension units, a Branch-Execution unit, two Integer units, one Floating-Point unit, one Load unit, and one Store unit. Nevertheless, even with the latter, the number of instructions that can be issued per cycle is still at most five, and the number of instructions that can be retired per cycle is still at most three.

1.6 SUMMARY

In this chapter we have introduced the fundamentals of pipelining. The basic ideas in the use of pipelining to obtain high performance have been discussed, and a taxonomy of pipelines has been given. Several measures for evaluating the performance of a pipelined machine have been introduced, and it has been shown that, ideally, the performance gains of pipelining are proportional to the degrees of pipelining and parallelism employed. We have also briefly discussed the factors that can prevent a pipeline from operating at its ideal performance; the detailed discussion of these factors, and how to deal with them, will make up most of the rest of the book. Lastly, some examples have been given of practical pipelines; these examples will serve as the basis of more detailed discussions in the chapters that follow.

2 TIMING AND CONTROL OF PIPELINES

In this chapter we shall look at various aspects of the timing and control of pipelines. An in-depth study of most of the topics covered here is outside the scope of the text, and what follows is only a brief introduction. The first section of the chapter consists of a derivation of approximate bounds on the pipeline cycle-time and a discussion of the effect of clock signals possibly not arriving as scheduled; as a consequence of the latter, we also, in the second section, look at the design of clock-distribution systems. The third section deals with the design of latches for pipelining — in particular with latches that combine storage and computational functions. The fourth section is an introduction to methods used in the detection of potential structural hazards and the corresponding design of pipeline-control.

2.1 CLOCK-CYCLE BOUNDS

The simple analysis in Chapter 1 suggests that we can arbitrarily increase the performance of a pipeline by increasing the number of stages employed while proportionately decreasing the cycle time. This is not possible, and in the first part of this section we shall briefly examine the basic constraints that bound the clock cycle time and, therefore, the number of stages permissible. (The number of stages is, of course, also constrained by the particular function to be partitioned and the requirement that the number be integral.)

A major issue in the timing is possibility that nominally identically timed signals may actually arrive at slightly different times: the *skew* is the difference in the arrival times of such signals at some point and may be deliberate or unintentional. Unintentional signal skewing is caused by a number of factors, of which physical layout, differences in lengths of propagation paths, and variations in loading are among the most common. Intentional clock skew is frequently used to avoid logic-hazards and in systems in which (essentially) several clock signals are employed (i.e. multi-phase clocking).

Ideally, the clock signal arrives at all pipeline stages at exactly the same moment. If we consider two adjacent stages of a pipeline (Figure 2.1) under the assumption that this is so, then we can deduce that (if data is staticized when the clock signal is high) the clock signal should be high for a duration that is long enough for the registers to operate properly, the clock pulse should not be high for a duration that would allow data to pass through more than one stage on the same cycle (a condition known as a *critical race*), and the clock period (i.e. the sum of the clock-high and clock-low times) should be large enough to allow for the proper operation of a stage (i.e. the latch plus subfunction logic).

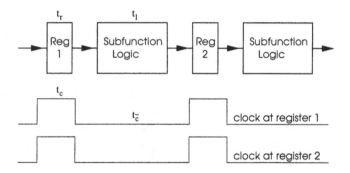

Figure 2.1.   Ideal clocking of pipeline stages

If we let t_r denote the time required for the operation of the register, t_l denote the time required by the subfunction logic, t_c denote the period when the clock pulse is high, and $t_{\bar{c}}$ denote the period when the clock pulse is low — that is, the clock cycle time is $t_c + t_{\bar{c}}$ — then the constraints above may be expressed as

$$t_c \geq t_r$$
$$t_c \leq t_r + t_l$$
$$t_c + t_{\bar{c}} \geq t_r + t_l$$

Note that the second relation ($t_l \geq t_c - t_r$) implies that there is a bound on how little logic a subfunction stage may have and implies that padding a subfunction unit with do-nothing logic may be necessary if the subfunction is nominally realized with fewer levels of logic. And the third bound implies that a reduction in the clock period may require a corresponding reduction in subfunction logic (i.e. an increase in pipeline length) or the use of faster registers, either of which might be impossible.

In practice, unless circuits are designed appropriately, physical and logical imperfections, as well as a number of other factors, can cause deviations from the ideal clocking depicted above. For example, clock pulses to adjacent stages may arrive at slightly different times — because, for example, they arrive via paths of different lengths or of different loadings. This phenomenon is known as *clock skew* and implies a need to refine to the bounds above.

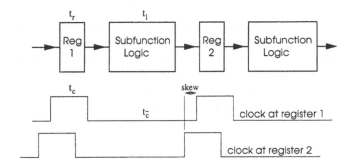

Figure 2.2.   Negative-skew clocking of pipeline stages

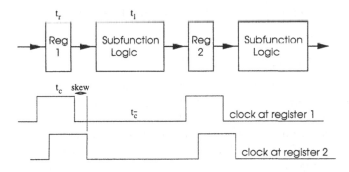

Figure 2.3.   Positive-skew clocking of pipeline stages

If we take the first of two adjacent stages, i and j, as a reference point, then we may say that the skew is *positive* or *negative*, according to whether the signal at stage j arrives earlier or later than that at stage i. The positive-skew case is shown in Figure 2.2. In this case, the time available for subfunction computation is reduced by the (maximum possible) clock-skew time, t_{sk}, and the time available for the register operation is also been similarly reduced; this affects the first and third bounds above and implies an increase by t_{sk} in the lower bounds on t_c and $t_c + t_{\bar{c}}$. If, on the other hand, we have negative skew, as shown in Figure 2.3, then the clock pulse has effectively been widened to $t_c + t_{sk}$; this affects the second bound above and increases, by t_{sk}, the lower bound on t_l. We therefore end up with the new bounds

$$
\begin{aligned}
t_c - t_{sk} &\geq t_r & (t_c \geq t_r + t_{sk}) \\
t_c + t_{sk} &\leq t_r + t_l & (t_l \geq t_c - t_r + t_{sk}) \\
(t_c - t_{sk}) + t_{\bar{c}} &\geq t_r + t_l & (t_c + t_{\bar{c}} \geq t_r + t_l + t_{sk})
\end{aligned}
$$

Allowing for for both positive and negative skew, the third bound should be replaced by

$$
t_c + t_{\bar{c}} \geq t_r + t_l + 2t_{sk}
$$

Furthermore, in practice individual gates can have different propagation times through them, resulting in *data skew*, so the constraints actually required are that sufficient time be allowed for the longest possible latch-operation time, that the effective clock pulse not be greater than the propagation time through the segment with the smallest delay, and that the effective clock pulse be long enough for the operation of the stage with the largest delay. These lead to the final bounds

$$
\begin{aligned}
t_c &\geq t_{r,max} + t_{sk} \\
t_c &\leq t_{r,min} + t_{l,min} - t_{sk} \\
t_c + t_{\bar{c}} &\geq t_{r,max} + t_{l,max} + 2t_{sk}
\end{aligned}
$$

Such bounds were first published by Cotten and later extended by others [2, 3, 7, 9, 12, 13].

2.2   CLOCK-SIGNAL DISTRIBUTION

Two of the basic issues in the distribution of clock signals are the provision of sufficient fan-out and the minimization of unintentional skew. There are a number of ways in which skew can be minimized, depending on its source: Skew arising from physical dimensions can be minimized by ensuring that the signal paths are of equal length (or as near to equal length as is possible), either by ensuring that interconnections are of the same length or by adding delays on what would otherwise be shorter paths; such an approach has been used in the Cray computers. Providing sufficient fan-out of the clock signal requires appropriate buffering, and this can also easily be combined with the minimization of path-length skew if all clock destinations are clocked from the same level of the clock buffer-tree. As examples, the clock distribution systems of the DEC NVAX and the DEC Alpha 21164 are shown in Figures 2.4, 2.5, and 2.6 [1, 6]; similar systems are used in a number of other current machines, e.g. the PowerPC [10].

The NVAX clock system minimizes path-length skews by placing the chip's global clock generator (which is driven by an off-chip oscillator) at the center of the chip and then radially distributing the clock signals. The outputs of the global generator go through four levels of inverters (to increase the driving ability) that in turn feed a central routing channel spanning the chip. The various functional units tap off this central channel, and additional buffering at the entry to each functional unit further increases the driving ability of the clock signals. Skew between global clock lines is reduced by adding dummy loads onto lightly loaded lines, thus improving the loading balance, and skew at the level of functional units is minimized by using a common buffer design and carefully tuning the buffer driver in each section. There is a final level of buffering just before the clock signals are finally used; this level of buffering reduces the load on the east-west routing lines and also sharpens the clock edges. The final clock skew is 0.5 ns, which is 4% of the 12ns cycle time.

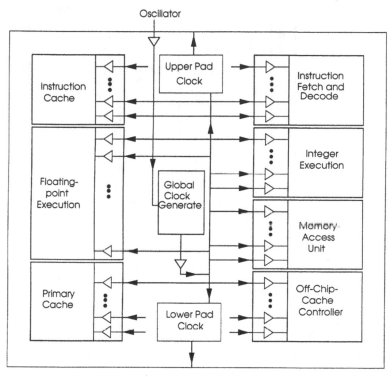

Figure 2.4.   NVAX clock distribution system

The DEC Alpha 21164 clock distribution system is similar to that of the DEC NVAX system in the central placement of the primary clock generator that is also fed by an external oscillator. The central generator feeds a fan-out tree of six inverter levels that in turn feed two sets of clock drivers, one on each side of the chip (Figures 2.5 and 2.6). Each of the driver sets consists of four more levels of inverter buffering, with a final set of forty-four drivers. There are also twelve conditional clock drivers on each side: one for each half of the second-level cache. The skew arising from the initial clock distribution tree (i.e. in the arrival of signals to the final sets of buffers) are minimized by strapping the inputs and outputs of the final clock drivers; and skew arising from differences in transistor characteristics of the final drivers are minimized by using a modular block for the final-driver layout. The final skew between the first and the last latch to receive the clock signal is 90ps, with a cycle time of 3.33ns. The complete clock distribution system accounts for 40% of the chip's power consumption.

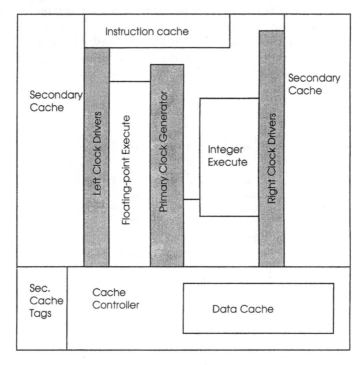

Figure 2.5.   Layout of Alpha 21164 chip

So far we have assumed that there is a single clock signal and that the objective is to deliver it simultaneously to all registers, whence the race condition above. Suppose, however, that the registers at the different stages are clocked with different signals, arranged in such a way that the registers at stage j are clocked only after those in register $j + 1$ have been. Then the race condition is eliminated, as the data output of one stage will be staticized only after those from the succeeding stage have been. Such a system is used in many machines, with the clock signals to different stages being derived from different phases of the same clock — *multi-phase clocking*. A common realization of this arrangement consists of using two phases, one for the even-numbered stages of the pipeline and one for the odd-numbered stages. Kunkel and Smith have shown that multiphase clocks can have shorter periods than single-phase clocks if the pipeline stages are long; this follows from the fact that with multiphase clocks and long segments, the clock phases can be adjusted so that the effect of data skew is reduced, whereas with short segments the effect of the data skew is comparable to that of the clock skew [12]. It should also be noted that the race problem could be solved by using registers constructed from flip-flops, i.e. latch-pairs in a master-slave arrangement, but these are more complex and slower than registers made from one-level latches.

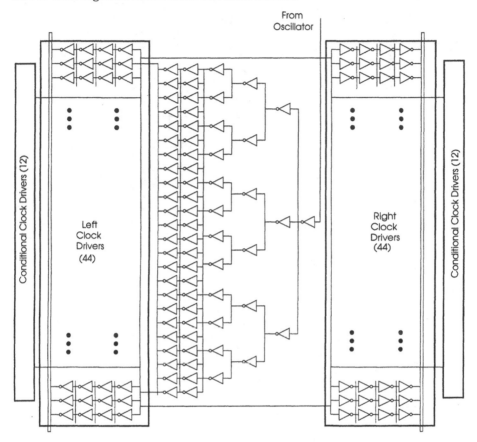

Figure 2.6.   Alpha 21164 clock distribution system

The clocking system used in the MU5 computer is an example of the type of system just described. Figure 2.7 shows the arrangement used for the pipeline of Figure 1.5: The pipeline beat starts at the last stage (when the highway has received the next order leaving the pipeline) and propagates backwards, with a delay at each stage. The end effect of this is that the signals received at the different stages are staggered as shown in Figure 2.8. (The shaded areas show the progress of one instruction through the seven pipeline stages.) Within each stage layout difficulties meant the registers that have to be clocked lie at varying distances from the delay chain, and this poses skewing problems. The solution here is to derive final clock signals from different levels of the fan-out structure, according to how far away the destinations lie — the signals for "far" registers are derived from earlier levels than those for "near" registers — and to design for only three levels of logic for data travelling from "near" to "far"

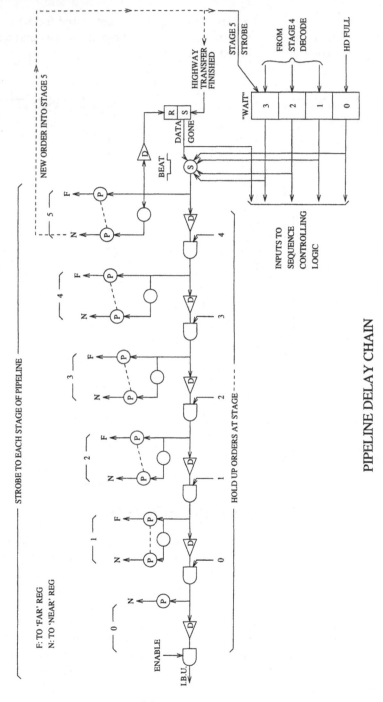

Figure 2.7.   MU5 clocking system

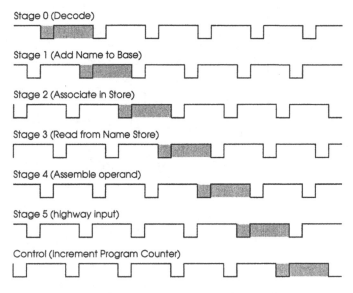

Stage 0 (Decode)

Stage 1 (Add Name to Base)

Stage 2 (Associate in Store)

Stage 3 (Read from Name Store)

Stage 4 (Assemble operand)

Stage 5 (highway input)

Control (Increment Program Counter)

Figure 2.8.   MU5 pipeline clocking and flow

registers. The other logic shown in Figure 2.7 is to generate pipeline stalls (hold-ups): to stop the part of the pipeline up to a given stage, the clock pulses are prevented from propagating beyond that stage.

2.3 LATCH DESIGN

A large number of high-performance arithmetic pipelines use carry-save adders (and other similar structures) that consist of just two levels of logic for the subfunction, and we have seen that a typical pipeline consists of alternating levels of subfunction-logic and latches. The *Earle latch* is a type of circuit that was designed specifically for such cases but which has since been used in a variety of pipeline structures [8, 13]. The latch allows the computation of a two-level AND-OR function (and therefore, in principle, any function of combinational logic) to be combined with latching in a manner that introduces no additional delay; that is, the resulting structure still has essentially a two-gate delay.

A one-bit Earle latch is shown in Figure 2.10. The stored information is whatever is on the DATA line when the CLOCK is a logical-1 and is retained as long as CLOCK is a logical-0; the value on the DATA line should therefore not be altered while CLOCK is a logical-1. The topmost AND-gate is used to enter new data, the bottommost AND-gate is used to maintain old data, and the middle one is used to eliminate logic hazards.

To see how the Earle latch might be used, consider the sum, S_i, and carry, C_i, bits in a full adder with operand bits A_i and B_i and a carry-in C_{i-1}. The relevant logic equations are $S_i = A_i B_i C_{i-1} + A_i \overline{B}_i \overline{C}_{i-1} + \overline{A}_i B_i \overline{C}_{i-1} + \overline{A}_i \overline{B}_i C_{i-1}$ and $C_i = A_i B_i + A_i C_{i-1} + B_i C_{i-1}$. If S_i and C_i are computed in one circuit

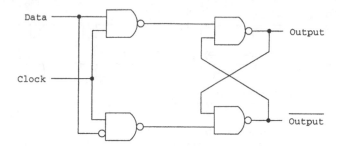

Figure 2.9.   One-bit D latch

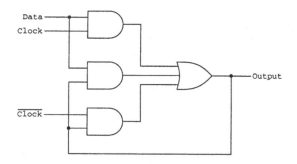

Figure 2.10.   1-bit Earle latch

and then stored in an ordinary latch (such as that in Figure 2.9), the delay will be 4τ (assuming that each gate has a delay of τ, that the inverses of all the signals are readily available, and that the latch used operates in a time of 2τ). It is, however, possible to use the Earle latch in the manner shown in Figures 2.11 and 2.12 to achieve the same effect within a time of 2τ; in essence, the propagation time through the latch is hidden. The circuits of Figures 2.11 and 2.12 have been obtained by substituting the definitions of S_i and C_i into the logic equations for the latch and then expanding: each of the first two gates in Figure 2.10 has now been replaced by four gates in Figure 2.11 and three gates in Figure 2.12.

Figures 2.11 and 2.12 show that ensuring that the circuits are hazard-free nearly doubles the logic costs (as measured by fan-in and fan-out or by the number of gates) and can easily lead to high demands in fan-in, fan-out, and clock distribution. The variant of the Earle latch obtained by omitting the middle gate in Figure 2.10 is known as the *polarity hold latch* and is commonly used instead [12]; as an example, the polarity-hold-latch analogue of the circuit in Figure 2.11 will have four fewer AND gates, with a corresponding reduction in the fan-out required of the clock and data signals and in the fan-in of the output OR gate. The design of the polarity hold latch does not eliminate hazards, but the latch will function correctly if $\overline{\text{CLOCK}}$ goes low just before CLOCK goes high; that is, if the signals are deliberately skewed.

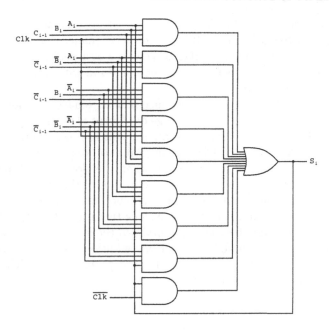

Figure 2.11.   Sum computation with Earle latch

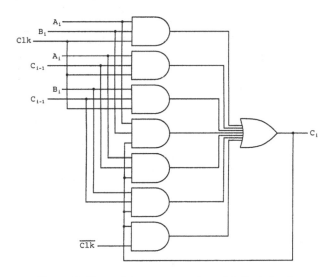

Figure 2.12.   Carry computation with Earle latch

Work by Kunkel and Smith has shown that Earle latch and the polarity hold latch have the same lower bound on the clock period [12]; the main difference, in terms of timing, between the the two types of latch is that the intentional-skew of the polarity hold latch imposes greater constraints on the permissible clock skew. The work referenced also has two additional main results: The first is that very short segments may have to be padded with intentional delay, and that if this is so, then wire pads will result in better performance than gate pads. And the second is that with no unintentional skew, the best performance is obtained with 6 to 8 levels of gating per pipeline segment; and with unintentional skew, 8 to 10 levels are required.

Designers of the latest high-performance machines have also developed specialized latches that take better advantage of current technology [18]. Thus, for example, the DEC Alpha 21164 uses latches that are built from transmission gates, as shown in Figure 2.13 [1]. (The top two latches are open when the clock is high, and the lower two are open when the clock is low: the chip uses a two-phase clock in which one phase is the inverse of the other.) To eliminate the possibility that data can race from one such latch, through subfunction logic, and into another latch on the same clock transition, the clock buffer that directly feeds into a latch is sized to keep down the skew, and the subfunction logic is designed so that there is at least one additional gate delay between all latches; the delay does not affect critical speed paths, all of which have more than one delay between latches.

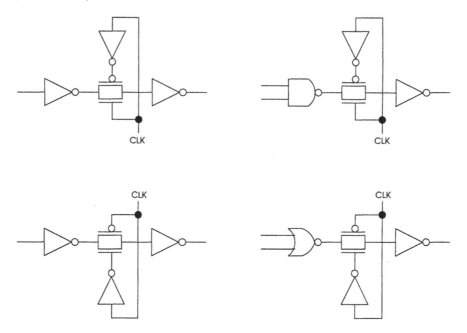

Figure 2.13.   DEC Alpha 21164 latches

To conclude this section, we remark that although pipelines have generally been implemented with inter-stage registers, it is possible (in theory, at least) to do away with these: The basic idea here is known as *wave pipelining* and was first suggested, by Cotten, in the 1960s [3]. Actually realizing this in practice has proved to be problematic, but after decades of neglect, the idea has recently been revived [11].

2.4   STRUCTURAL HAZARDS AND SEQUENCING CONTROL

In this section we consider the problem of ensuring that a pipeline can attain the highest throughput even while avoiding structural hazards; that is, how to control the pipeline so that new orders will be introduced into the pipeline at the highest rate that can be achieved without having collisions in the use of shared resources. The two main issues to be dealt with are the detection of possible structural hazards and the determination of the *initiation latency*, which is the the minimum time must elapse between the initiation of one order and that of the next order. The basic algorithm for both was developed by Davidson and later extended by others [5]. We next describe this algorithm, initially for a unifunctional pipeline and then for a multifunctional one.

The flow of orders through a pipeline, with an indication of resource usage, may be represented by a *reservation table*, a two-dimensional table in which the rows correspond to the pipeline stages and the columns correspond to time (cycle) units. Slot (i, j) of the table is marked if after j cycles the order initiated in cycle 1 requires the use of stage i of the pipeline. Consider, for example, a unifunctional pipeline (based on Figure 1.1) in which a single adder is used both in the first stage to increment the program counter and in the fourth stage to execute ADD instructions (Figure 2.14). The (initial) corresponding reservation table is shown in Figure 2.15(a). To determine the intervals at which it is safe to initiate new orders, a copy of the reservation table is overlaid on itself, with a shift of k time units, for different values of k. If there is no overlap of the markings, then it is safe to initiate a new order k time units after the initiation of the last order; otherwise, it is not safe, and an interval of more than k time units must elapse before it is safe to do so. For the example of Figure 2.15(a), the overlay for $k = 1$ is shown in Figure 2.15(b); there is no overlap there. Similarly, there is no overlap with $k = 2$. With $k = 3$, however, there is an overlap, as shown in Figure 2.15(c), which indicates a potential hazard; this is what we should expect, as an ADD instruction initiated in cycle 1 will be using the adder in cycle 4 (its EXECUTE phase) and at the same time a new order initiated in that cycle will also require the adder to alter the program counter. The value of k for which a collision occurs is known as a *forbidden latency*, and the list of all forbidden latencies is the *forbidden list*. The forbidden list for the pipeline of Figure 2.14 is (3).

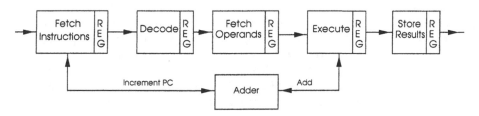

Figure 2.14.   Pipeline with shared resource

Once the forbidden list has been determined, it is used to construct a *collision vector*, from which a hardware control structure is derived: If n is the largest value in the forbidden list, then the corresponding collision vector is an n-bit binary vector, $c_1 c_2 \cdots c_n$, where c_i is 1 if i is in the forbidden list and 0 otherwise.. The collision vector is so-called because it indicates when collisions can occur: a 1 in position i of the vector indicates that initiating a new order i cycles after the last one was initiated will result in a collision, whereas a 0 indicates that a new order may safely be initiated. Thus, for example, the collision vector for the pipeline of Figure 2.14 is 001, which indicates that a new order may be initiated in the first two cycles after the last one was, but not in the third cycle after. The collision vector is then used by implementing a hardware structure that checks for 0s in all positions that correspond to the number of cycles that have elapsed since the last order was initiated. Because of its simplicity, a shift register is a suitable structure and is used as follows. The register is initialized to the initial collision vector. At every cycle thereafter, the contents of the register are shifted one position to the left. If the bit shifted out is a 1, then no new order is initiated, and the contents of the register are changed to the logical OR of each bit of the register and the corresponding bit of the initial collision vector; otherwise, a new order may be initiated.

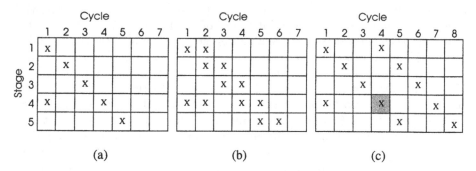

Figure 2.15.   Reservation tables

The basic technique just described above is easily extendible to a multifunctional pipeline: A reservation table is constructed for each type of order that the pipeline may process. Then, overlaying one type of table onto itself will determine the forbidden latencies when an order of that type follows an order of the same type, and overlaying one type of table onto a different type of table will determine when an order of the first type may safely follow an order of the second type. A collision vector is then constructed for each forbidden list, and the implementation of the control is as for the unifunctional pipeline, except that the OR operation will now involve all collision vectors. We leave it as an exercise for the reader to try this on, say, the TI-ASC pipeline described in Chapter 1.

2.5 SUMMARY

In this chapter we have briefly discussed the factors that determine lower bounds on the clock cycle of a pipeline and, by implication, an upper bound on the number of stages. The most important of these factors is clock skew, and methods for distributing the clock signal in a way that minimizes skew have also been examined. The design of latches for high-performance pipelines has also been discussed; there are a variety of such latches, all of which are designed to make the best use of current technology, but their detailed design is outside the scope of this book. Lastly, we have considered structural hazards and how to design for proper sequencing their presence. In current high-performance machines the availability of resources is hardly a major problem, and the greater challenge is how to effectively utilize resources, rather than how to deal with competition; this explains the relatively little space devoted above to structural hazards.

3 HIGH-PERFORMANCE MEMORY SYSTEMS

A high-performance pipelined machine requires a memory system of equally high performance if the processor is to be fully utilized. This chapter deals with the issue of matching processor and memory performances. In a simple machine in which a single main memory module is connected directly to a processor (with no intermediate storage), the effective rate at which operands and instructions can be processed is limited by the rate at which the memory unit can deliver them. But in a high-performance machine, the access time of a single main memory unit typically exceeds the cycle time of the processor by a large margin, and obtaining the highest performance possible requires some changes in the basic design of the memory system. The development of faster memory as a solution to problem is not viable due to limitations in the technology that is available within a given period: the performance of the technology used for main memories typically improves at a rate that is less than that used processors, and the use for main memory of the fastest available logic is never cost-effective.

There are two main ways in which memory systems are usually designed so that they can match the rate at which a processor issues requests. The first if the use of (smaller) faster intermediate storage, between memory and processor, to hold temporary results, frequently used operands and instructions, and prefetched operands and instructions; this intermediate storage usually consists of a *cache*, or a set of registers, or some other type of buffering. The second is to arrange the memory system so that several words can be accessed in each memory cycle. This is usually done by replacing a single memory unit by several memory units (*banks*), organized in such a way that independent and concurrent access to several units is possible. The technique is known as *memory interleaving*, and it works best with predictable memory-access patterns. The first approach aims to reduce the *effective* access time of the memory by reducing the number of accesses that actually reach the memory: by arranging, whenever possible, for data to be frequently found in faster storage, the overall effect is that for most of the time the processor is not held up by the low speed of the main memory and operates as though the main memory was much faster. The second approach reduces the effective access time by increasing the memory bandwidth. In this chapter we shall examine these two techniques in more detail. The first part of the chapter section deals with memory interleaving, the second deals with caches, and the third is a summary.

3.1 MEMORY INTERLEAVING

Suppose we have a simple processor with a cycle time of τ_p directly connected to a single memory bank with an access time of τ_m and that τ_m is much greater than τ_p. Then the processor will be greatly slowed down by the memory unit. If, however, we could arrange N memory banks, where $N = \lceil \tau_m/\tau_p \rceil$, in such a way that all could be accessed in parallel in the same time as it would take to access a single bank, then, since N words would be delivered at a time, the memory access time, as seen by the processor — that is, the effective memory access time — would be $\lceil \tau_m/N \rceil$ and would therefore match processor cycle-time. This arrangement is what takes place in memory interleaving, and in many practical implementations interleaved memories supply data at a rate that exceeds the nominal processor requirements; that is, N is larger than $\lceil \tau_m/\tau_p \rceil$. Interleaved memories are especially useful in vector-processor machines, in which the sizes of data sets make caches less effective than usual and the memory system has to be capable of supplying the processor with a continuous stream of data at a high rate.

Another advantage of having multiple banks in a memory system is that it increases fault-tolerance: With a single memory unit, operational failure can be catastrophic, but with multiple banks *fail-soft logic* can be incorporated to reconfigure the remaining banks into a smaller memory for normal operation. Thus, for example, with a 4K four-bank memory system, the failure of one bank would cause reconfiguration to a 3K memory, failure of two banks to a 2K memory, and so forth.

The first part of this section covers various basic aspects of interleaved memories, the second deals with addressing patterns, and the third consists of case studies.

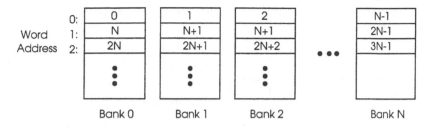

Figure 3.1.   Assignment of addresses in memory interleaving

3.1.1 Basic principles

For the interleaving to work as indicated, it is necessary that every N temporally adjacent words[1] accessed by the processor be in different banks. If we assume that processor accesses are to sequential addresses — and this is generally true for most accesses to instructions and to structured data, such as vectors — then this requirement is easily met by placing assigning a memory address k to memory bank $k \bmod N$. This gives the arrangement shown in Figure 3.1. (In general, the interleaving may be done on the basis of a unit that is smaller or larger than a single word; see Section 3.1.3 for an example). To access a memory location, the memory system interprets each processor-generated address as consisting of a pair $\langle bank\_address,\ displacement\_in\_bank \rangle$; that is, for a memory of K words interleaved in N banks, the low-order $\log_2 N$ bits[2] of the address are used to select a bank and the high-order $\log_2 K - \log_2 N$ bits are used to select a word within the selected bank[3].

Ideally, an N-way interleaved memory will have a bandwidth of N words per memory cycle. But this will be so only if every N addresses delivered to the memory system are independent, as indicated above, and this is not always the case: in practice, we should expect that branches in the instruction stream and unpredictable accesses in the data stream will result in an overall performance that is less than this maximum. It is, however, possible to buffer processor requests and then service them in an order that maximizes the performance of the memory system, as opposed to servicing them in order of arrival. Empirical observation and analytical studies both indicate that if requests are to randomly distributed addresses and those to busy banks are not buffered, but are serviced in order of arrival and therefore allowed to block other requests, then the effective bandwidth for an N-bank memory will only be of the order of $\sqrt{N}$ words per cycle [5]. Many implementations are therefore organized so

---

[1] The interleaving may be on the basis of any addressable unit, which may be smaller than a word, but for simplicity we shall assume word-interleaving, which is the most common.

[2] We are here assuming that N is a power of two, and it usually is. There are, however, advantages in having N take some other values (Section 3.1.2).

[3] This is known as *low-order interleaving*; *high-order interleaving*, the alternative, is not useful for increasing bandwidth but allows for easier expansion of memory.

that the memory system can accept one request in each processor cycle if the addressed bank is not busy.

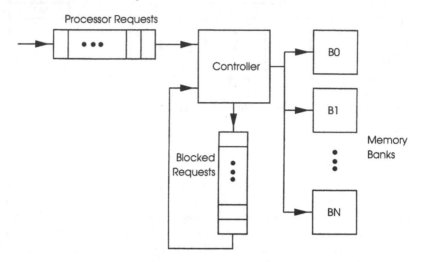

Figure 3.2.   Interleaved memory system with blockage- buffer

The high-level organization of a memory system with blockage-buffering is shown in Figure 3.2. The main components are the memory banks, a queue that buffers requests from the processors, a queue[4] for requests that are held up in busy banks, and a controller to coordinate the operation of the system. In each (processor) cycle, the controller takes a request from either of the queues, giving priority to the blocked-requests queue, and either forwards it to the addressed bank (if the bank is not busy) or enters it at the tail of the blocked-requests queue (if the bank is busy). This buffering has a number of important implications: First, since the order in which the requests are serviced at the banks is determined by the order in which they arrive at the memory system, the addresses involved, and the order in which banks become free or busy, it follows that the order in which the processor receives responses can be different from the ordering of the corresponding requests; so some means of matching responses with request-addresses may be necessary. Second, and partly as a consequence of the preceding, the controller must ensure consistency between the responses and the requests; for example, if a LOAD (READ) request is blocked, then a following STORE (WRITE) that happens to find free the bank addressed by the LOAD should not always be allowed to proceed, since it may be an access to the location to which the LOAD is directed; this may be refined to blocking such a STORE only if is to the same location, but the implementations costs will be higher. Third, the arrangement allows certain performance-enhancing refinements: for example, two LOADs to the same mem-

---

[4]In actual implementation, there may be a queue for each bank, rather than a single shared queue as shown.

ory location (if they occur within a suitable time interval) can be satisfied with a single request to the memory bank concerned; a LOAD that follows a blocked STORE to the same location can be satisfied without accessing memory; and two blocked STOREs to the same location can, depending on whether or not there are any intervening LOADs to the same location and how these are handled, be collapsed into a single STORE. (As we shall see below, similar optimizations can also be integrated into a cache system.)

Three examples of the type of systems just described are given below as case studies. In one of these the blockage buffer is a single set of registers and there is no attempt to include the performance optimizations described above; in another, the blockage buffer is implemented as several structures, in an arrangement that the realizes the optimizations described above.

3.1.2 Addressing patterns

For the address-to-bank (sequential) assignment shown in Figure 3.1, if the data to be accessed is stored sequentially (i.e. corresponding to the addresses), then the type of reference pattern that results in the best performance is one in which the differences between two consecutive addresses is 1 — this is commonly referred to as *unit stride* — with a non-unit constant value as the next best. In practice, however, the stride is not always 1, even for a single data structure: consider, for example, a matrix that is laid out by rows across several memory banks (Figure 3.3) and then accessed, in order, first by rows and then by columns; in this case unit-stride exists for row accesses, but not for column or diagonal access. Evidently, the problem is worse for two-dimensional structures than for one-dimensional ones; but even with the latter, the need to extract substructures can result in irregular stride (see Chapter 6).

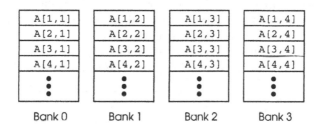

Figure 3.3.   A matrix layout in interleaved memory

The effect of stride is determined by two main factors: the behaviour of the program being run and the layout of data within the memory banks; the latter determines whether or not consecutive references are to independent banks, regardless of stride. Consequently, there are three main techniques that may be used to optimize the performance of an interleaved-memory system: the first is to modify programs so that they produce smaller strides than would otherwise be the case (i.e. with unoptimized code), the second is to store data structures in a way that increases the probability that consecutive accesses in a typical address-pattern will be to independent banks, and the third is to interleave (i.e. assign addresses to banks) in a manner that tries to ensure that consecutive addresses are to different banks, though not necessarily with unit stride. As we might expect, only limited success can be achieved with the first approach, although some non-trivial examples are given in Chapter 6. The other two approaches are generally more profitable, and we next look at some techniques that have been used to realize them.

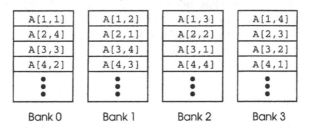

Figure 3.4.   Skewed addressing in interleaved memory

Skewed addressing is a common organization used to store matrices in a manner that increases the probability of parallel access to its components. The arrangement consists of cyclically shifting the starting point for each row, by a specified distance (the *skew*). As an example, Figure 3.4 shows a four-by-four matrix, A, laid out in four banks with a skew of one. In contrast with the situation described above, it is now possible to have parallel access to elements of a row or elements of a column, although not to elements of a diagonal. If the number of banks employed is prime, then skewing will also yield parallel access to elements of diagonals. An example is shown in Figure 3.5. Such a scheme has been implemented in the Burroughs BSP machine, with 17 banks [18]. In general, a memory system with straightforward interleaving will have good performance when the stride is prime relative to degree of interleaving; if this is not the case, then some references will be to to the same bank, in smaller reference-intervals, with correspondingly poor performance.

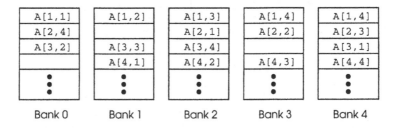

Figure 3.5.   Prime number of banks in interleaved memory

An alternative to using a prime number of banks is to interleave on a random (or pseudo-random) basis, the essential idea being that randomization will tend to uniformly spread the references between the banks and therefore increase the likelihood of bank-independence for consecutive accesses. The process consists of applying a "randomization" function to each address and then using the result to access the memory. There is, however, a cost to random-skewing: the length of the blocked-requests queue tends to be longer than usual during high-rate accesses, and, therefore, there is an increase in the average access time. The Cydra 5 is an example of a machine in which this has been implemented [25, 26].

3.1.3   Case studies

We now look at three case studies of the implementation of interleaved memory system. The first is the memory system of the CDC 6600, which is at the root of the family-tree of supercomputers from CDC and Cray Research; it exemplifies straightforward interleaving with blocked-request buffering [32]. The second case study is the implementation of the memory system in IBM 360/91 [3]. This implementation shows a blocked-request-buffering system that has been organized to give the best performance on sequences where some LOAD/STORE requests are to the same locations; it contains a number of the essential ideas used in the implementation of many of today's high-performance microprocessors. The third case study is the system used in the Cyber 205 [6]; this is typical of the implementations used in current vector supercomputers, which we shall study in Chapter 6.

3.1.3.1   CDC 6600 Stunt Box. The CDC 6600 central processor has a cycle time that is one-tenth that of the memory unit, and interleaving is therefore used to match the disparate rates. The memory unit is interleaved into several banks and is controlled by a unit known as the Stunt Box.

The Stunt Box has the organization shown in Figure 3.6. The three main parts are the Hopper (which comprises the registers M1, M2, M3, and M4), the Tag Generator, and the Priority Network. The Hopper serves as a staging post for addresses that are temporarily held up on a busy bank, and the Priority Network determines which of three possible sources (the central processor, a peripheral processor, and the Hopper) of a memory address gets to use the bus in the event of concurrent requests. Central processor requests come from two sources — one for an instruction and one for an operand — and the Priority Network resolves any conflict between the two by allowing operand requests to proceed first. The Priority Network also filters out illegal addresses and ensures that data consistency is maintained in successive LOAD/STORE sequences to the same memory locations. The Tag Generator tags each address according to the type of request; there are twenty-seven different tags.

The operation of the Stunt Box is as follows. Each incoming CPU address is entered in the register MO and interpreted as a displacement into the memory partition allocated to the current program: the RA register holds the base address of the partition and another register, FL, holds its length. The contents of MO and RA are then added together to obtain an absolute memory address. Concurrent with the addition, a check is made, by comparing the contents of

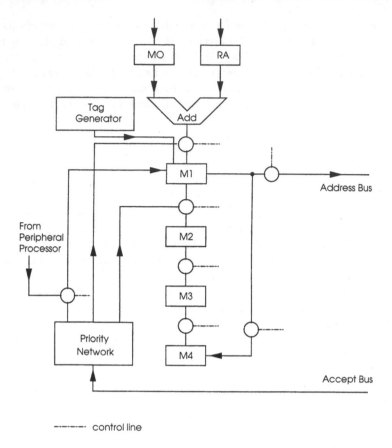

Figure 3.6.   CDC 6600 Stunt Box

MO and FL, to ensure that the address falls within the allocated partition. If the address passes this test and there is no waiting address in M2, then it is tagged, entered in the register M1, and placed on the address bus a short while later. The address then progresses through the registers M4, M3, M2, and possibly back to M1, at intervals that are equal to three-fourths of the processor cycle time. Just before the address is due to be transferred back into M1, the memory bank that is addressed sends back a response, on the Accept Bus, indicating that the request has been accepted or that the bank is busy. If the bank is not busy, then the address is removed from the Hopper (by disabling the transfer from M2 into M1); otherwise it is entered into M1, with priority given to a recirculating address over any incoming requests from the central processor or from a peripheral processor. It is, however, possible that while an address is in, say, M4, another request to the same bank will come in and, find the bank free, and be immediately accepted, thus causing further delays to the address in the Hopper. But this cannot continue indefinitely: it can be shown that the worst possible delay in such a case is two memory cycles.

3.1.3.2   IBM 360/91 Main Storage Control Element.

The main memory on the IBM 360/91 has an access time that is ten times the processor cycle time and consists of 4, 8, or 16 banks, depending on the particular machine configuration. Access to the banks is controlled by the Main Storage Control Element, which has the organization shown in Figure 3.7. This unit processes requests that each consists of the address of the memory location to be accessed, an operation specification (LOAD or STORE), and a sink address (for a LOAD operation) or data (for a STORE operation). A request is forwarded to the bank addressed if that bank is not busy and buffered otherwise. The buffering is arranged to allow several requests to be outstanding, for requests to be serviced out of their initiation order whenever doing so does not compromise consistency, and for requests to be serviced without accessing a bank if possible. That is, all of the optimizations described above are implemented.

The primary components of the Main Storage Control Element are the Store Address Registers, which hold addresses of data to be written into memory; the Store Data Buffers, which hold data waiting to be written into memory; the Request stack, a set of four registers, organized as a first-in-first-out queue, that buffers memory requests that have been rejected because of busy banks; the Accept Stack, a set of registers, organized as a first-in-first-out queue, that holds information on requests that have been accepted and are being processed; the Storage Address Bus, which transmits addresses to the banks; the Sink Address Return bus, which, for memory fetches, transmits register-sink addresses to sink registers one cycle before the data fetched is returned[5]; the Storage Bus Out, which transmits data from the memory banks; the Storage Bus In, which transmits data from the Store Data Buffers to the memory banks; and the Protect Memory, which holds keys required to ensure protection during address translation.

For a LOAD operation the operation of the Main Storage Control Element starts with the placement of the address on the Storage Address Bus. If the addressed bank is not busy, then the address is compared with all entries in the Storage Address Registers; otherwise, it is entered in the Accept Stack and also compared with the entries in the Request Stack. A match with an entry in the Storage Address Registers means that the addressed word is the target of an uncompleted STORE operation; to ensure that the correct data is returned, the request is entered in the Request Stack until the STORE has been completed. If there is no matching entry in the Storage Address Registers, then the request is sent to the addressed memory bank, and an appropriate entry is made into the Accept Stack; the entry consists of the address of the memory location, the address of the bank, and the register-sink address. A match with an entry in the Request Stack means there is already a waiting request to the same address; in this case the current request is entered in the stack, after the older one to ensure proper sequencing. If, however, the comparison with the entries in the

---

[5]LOAD requests may be serviced out of their initiation order, and a LOAD request may nominally be satisfied before the data is returned.

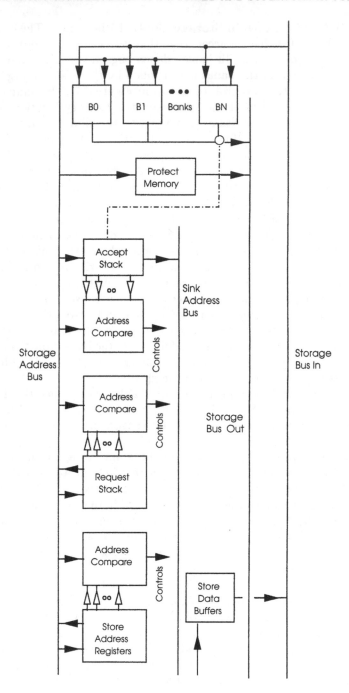

Figure 3.7.   IBM 369/91 Main Memory Control Element

Request Stack is unsuccessful, then a comparison with the entries in the Accept Stack is carried out. A match with an entry in the Accept Stack means that a LOAD request to the same memory location is already being processed by the bank. If the earlier request is a LOAD, then the present request can be satisfied at the same time as the earlier one and without accessing the addressed bank; and if the earlier request is a STORE, then the present one can be satisfied from the Store Data Buffers, again eliminating the need to access the bank concerned. Here too, an unsuccessful comparison comparison with the entries in the Request Stack causes the address to be entered in the Request Stack.

3.1.3.3 CDC Cyber 205 Priority Unit. The central processor of the CDC Cyber 205 operates on a cycle that is one-fourth that of the memory. To balance the two rates, a highly interleaved memory is used: The memory on a fully configured Cyber 205 is arranged in thirty-two *superbanks* (or *modules* in CDC terminology), with a 32-bit *half-word* as the basic addressable unit. Each superbank consists of eight banks. The banks are independent, and each bank may start a new cycle of operation before adjacent banks in the same superbank have completed their operations; however, only one bank in the same superbank may start operation in the same processor cycle, and no bank may accept more than one request in the same memory cycle. Data may be transferred in units of half-words, words, 512-bit *super-words* (*swords*), or 1024-bit double-swords. Since the banks are interleaved on a half-word basis, a single memory reference may require accesses to more than one bank: word, sword, and double-sword references require access to two, sixteen, and thirty-two banks, respectively. During most vector operations, the memory system operates in in *stream mode*, in which a half-word is accessed in all thirty-two superbanks, yielding a double-sword in each memory cycle.

Memory accesses in the machine are controlled by a Priority Unit that is similar to the CDC 6600 Stunt Box. In its operation, the Priority Unit first discards any illegal addresses and then arbitrates among multiple requests if there are any. If a request has sufficiently high priority, the address is an absolute address or a virtual address for which there is a currently-mapped translation register, and the addressed bank is not busy, then the memory operation is initiated and an *accept* signal is sent back to the source. Otherwise, a translation register is loaded or Retry Unit that is similar to the CDC 6600 Hopper (with buffers to hold up to four waiting requests) repeats the request.

The requests that arrive at the Priority Unit are divided into two classes: *immediate-issue* and *delayed-issue*. The former consists of all LOAD requests and STORE requests involving a unit smaller a sword; the latter consists of all STORE requests of a sword or double-sword. The main reason for this separation is that delayed-issue requests require a sufficient delay to permit the accumulation of data into sufficiently large units in the memory data buffers. When a delayed-issue request is accepted, the addressed banks are reserved for three processor cycles, since the memory cycle is equal to four processor cycles; no other request is accepted during this period if it is to one of these banks.

For each request it receives, the Priority Unit also carries out two tests on the status of each bank addressed: a *pre-issue check* and a *post-issue check*. The pre-issue check consists of comparing the address of the bank required by an immediate-issue request with the addresses reserved by delayed-issue requests in progress; if a match occurs, then the immediate-issue request is held up. The post-issue check consists of determining whether or not a request of the same type as the present one has referenced the addressed bank within a period of three processor cycles; if that is the case, then the present request is held up. An immediate-issue request also gets held up if it follows a delayed-issue request within a period of four processor cycles; this hold-up is necessary because there is only one address bus to the banks.

3.2 CACHES

A cache is essentially a high-speed, (usually, but not always) programmer-invisible[6] buffer between main memory and the central processing unit. The cache is intended to take advantage of program-referencing patterns and thus provide, as far as seen by the processor, a memory system that is fast and large but which is reasonably cheap. The basic idea is that provided most of a program's memory references can be satisfied from high-speed memory, the relatively large access time of the main memory need not be visible to the processor. The effectiveness of this relies on the property of *locality*, of which there are two forms: *spatial locality* (i.e. locality in space), which means that if a reference has just been made to a particular memory location, then it is likely that subsequent references will be to adjacent memory locations, and *temporal locality* (i.e. locality in time), which means that if a particular memory location has just been referenced, then it is likely that the same memory location will be referenced again in the near future. (As an example, the execution of a program that adds two vectors will exhibit high spatial locality over the vectors, and that of a program that forms a vector dot-product will exhibit high temporal locality in the accumulation variable.) So, provided data are kept in a high-speed buffer, according to use (by recency or location), access to main memory need not be frequent. A single cache may be used for data only, or for instructions only, or for both instructions and data; in the last case the cache is said to be *unified*, and in the other two cases a two-cache system is said to be *split*.

In the general organization of a typical cache-based memory system, the main memory is divided into *blocks*, each of several words[7], the cache is divided into *lines* (or *block-frames*) of the same size as the blocks, and the blocks are mapped, as needed, into the lines according to some predetermined *placement policy*. Each line of the cache has one or more associated *validity* bits that at any given moment are set or not, according to whether or not the line contains valid data. All processor references to memory first go though the cache. If a

---

[6]Caches in the some machines allow limited programmer/compiler control.
[7]We shall assume that the basic addressable unit is a word, although in practice it could be of any size.

required data is found in the cache — in this case a *hit* is said to have occurred — then it is returned to the processor in a time determined by the cache access time. And if the data is not in the cache — in this case a *miss* is said to have occurred — then the block containing it is brought into the cache and the datum is then sent to the processor; the time taken in this case is determined by both the cache access time and the main memory access time. A block may also be fetched from the main memory at other times, depending on the *fetch policy* employed.

During the normal course of a cache's operation, it will sometimes be necessary to eject a block from a cache line in order to make room for a new block from main memory. In such a case a *replacement policy* determines which block is to be ejected. Lastly, in the case of a WRITE (STORE) operation into the cache, the *write policy* determines when the corresponding memory location is updated.

3.2.1 Placement policies

The most restrictive placement policy is *direct-mapping*, in which a given block can be placed in only one specific line of the cache. The address of the line is usually calculated as the block address modulo the number of cache lines, this being an especially simple function to implement in hardware. In order to determine which block occupies a line at any given moment, each line has a *tag* field that identifies the block it contains. Because a number of blocks, with the same low-order address bits, are thus mapped onto the same line, the tag holds only the high-order address bits of the block. Thus, for example, with a main memory of 1024 blocks and a cache of 32 lines, the mapping and the partitioning of a memory address would be as shown in Figure 3.8.

For a LOAD (READ) access, each memory address from the processor is then processed as follows. First, the line onto which the block is mapped is identified by applying the placement function. For an address of k bits, a line of m words, and a cache of n lines, this involves dropping the low-order $\log_2 m$ address bits (i.e. those that identify the word within a line), dropping the high-order $k - \log_2 n$ bits of the address (i.e. those that differentiate between blocks mapped onto the same line) and using the remaining $\log_2 n$ bits to select a line of the cache. Next, the high-order $k - \log_2 n$ bits of the address are compared with the tag in the addressed line. If there is a match, then the low-order $\log_2 m$ address bits are used to select a word of the line, and the validity bits are then examined to determine whether or not the selected data in the line is current; current data is returned to the processor. And if there is no match or current data, then an access is made to the main memory and the line updated when the data are returned. The complete accessing arrangement , then, is as shown in Figure 3.9. For a STORE (WRITE) access, the address is processed as for a LOAD, and then if the block to be written into is in the cache, then the data is written in as the last step; if the block is not in the cache, then subsequent actions depend on the write policy (Section 3.2.2)

In practice, the above scheme is usually implemented so that both the comparison with the tag and the data-readout can proceed in parallel, with subsequent cancellation there is no hit (see Figure 3.9). However, for STOREs, the write policy may not always allow concurrent tag-check and writing (Section 3.2.2).

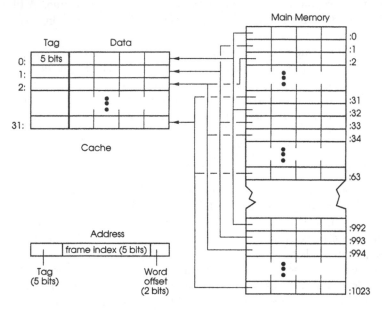

Figure 3.8.   Mapping in a direct-mapped cache

The least restrictive placement policy is *fully-associative mapping*, in which a block can be placed in any line of the cache. Thus, for example, for a cache of 32 lines and a main memory of 1024 blocks, the mapping and partitioning of the address would be as shown in Figure 3.10. In order to determine what block is mapped onto a line at any given moment, the tag field must now hold the full address of the block, and a comparison must now be made between the tag and all of the bits of incoming memory address, except for the word-identifier bits. Because the comparison has to be made against all the tags of the cache and has to be fast, the tags are held in associative (content-addressable) memory, which allows concurrent comparison and returns the address of the line that gives a match and has valid data. The processing for a LOAD operation is depicted in Figure 3.11; for a STORE, similar remarks as those for the direct-mapped cache apply. Note that unlike the direct-mapped cache, here tag-checking and data-readout cannot proceed in parallel, as the line from which the data is to be read is not known until the tag-check is complete.

In between the two extremes of the direct-mapped cache and the fully-associative cache is the *set-associative cache*, in which a block can be placed in just one specific *set* (of several lines) of the cache; a cache with n lines per set is said to be n-way set-associative. As an example, with 32 cache lines, 4 lines

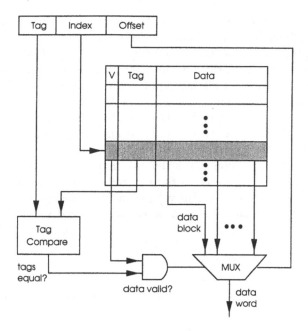

Figure 3.9.   LOAD access in a direct-mapped cache

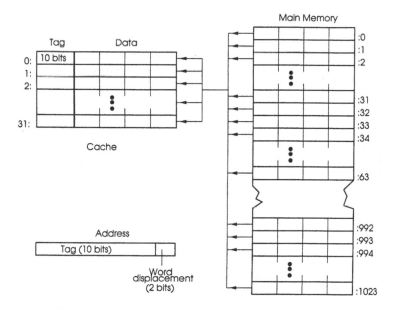

Figure 3.10.   Mapping in fully-associative cache

per set, and a 1024-block memory, the mapping and partitioning of the memory address would be as shown in Figure 3.12. Access in a set-associative cache first proceeds as in a direct-mapped cache and then as in a fully-associative cache: First, the set whose address is obtained by applying a direct-mapping function to the incoming block-address is selected. Then the high-order bits of the incoming block-address (i.e. those bits obtained by excluding the bits that identify the set and the word) are concurrently compared with all the tags in the selected set. Lastly, for a LOAD a word is selected from the line that gives a match and has a set validity bit; and for a STORE, the situation is the same as in the other two types of cache. Data-readout and tag-checking cannot be carried out in parallel in a set-associative cache, for the same reasons as in a fully-associative cache.

The preceding description shows that we may view a direct-mapped cache as a set-associative cache in which the number of sets is equal to the number of lines (i.e. as 1-way set-associative) and a fully-associative cache as a set-associative cache with one set (i.e. as k-way set-associative for a k-line cache).

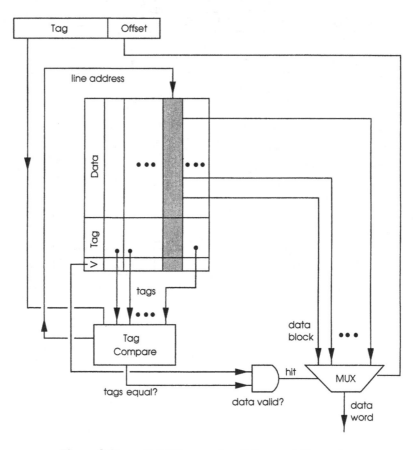

Figure 3.11.   STORE access in a fully-associative cache

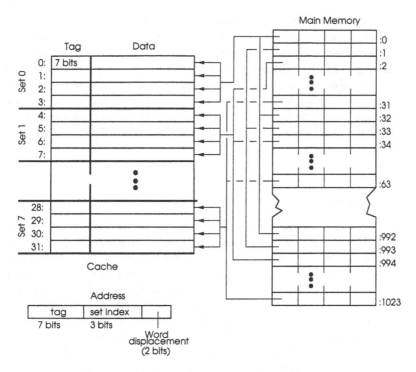

Figure 3.12.   Mapping in a set-associative cache

3.2.2 *Replacement policies*

The replacement policy for a direct-mapped cache is simple and fixed: if a fetched block maps onto an occupied line, then the contents of that line are replaced. In the case of a fully-associative cache and a set-associative cache, there are a number of candidates for replacement algorithm: for example, with *least-recently-used* (LRU), the block selected in the cache or set is the one that has least recently been accessed for reading or writing; with *random* (RAND), a randomly selected block is replaced; and with *first-in/first-out* (FIFO), the block that was brought in earliest is replaced.

RAND is simple to implement in hardware, spreads allocation uniformly, and results in easily reproducible patterns, which is useful for debugging, but it does not take program behaviour into account and therefore cannot be expected to give the best performance. Example implementations of RAND are in the data cache of the Motorola 88110 (Section 3.2.5.3), one of the caches in the DEC Alpha 21164 (Section 3.2.6.2), and in the Branch-Target Cache (a special type of cache for handling branch instructions) of the Am29000 (Chapter 4). FIFO is harder to implement in hardware than RAND but easier to implement than LRU; examples of machines that implement FIFO are the AMD K6, Hitachi Gmicro/200, and the DEC VAX 9000 [1, 9, 13]. LRU is intuitively the best policy, as it takes program behaviour into account — programs tend to make repeated accesses to the same memory locations, which is one reason why a cache is useful in the first place — but is relative costly to implement, especially if the set (in a set-associative cache) or the cache (fully-associative cache) is large. Many implementations therefore use only an approximation to LRU. The most common such approximation is *not-recently-used* (NRU), which simply selects a block that has not been used recently; an example of an NRU implementation is described in Section 3.2.5.1 An example of an LRU implementation will be found in the instruction cache of the AMD K6; this machine also implements a relatively unusual algorithm for the data cache – the *least recently allocated*.

3.2.3   Fetch policies

The *fetch policy* determines when a block is to be brought into the cache. The simplest policy is *demand fetching*, in which a block is fetched only when a miss occurs. On the other hand, a *prefetching* policy tries to reduce the number misses by using some prediction to fetch blocks before they are actually referenced [12, 29, 31]. One such prediction is that if block i is referenced, then block $i+1$ will be referenced in the near future, and so it too should be brought into the cache if it is not already there. This strategy, which is known as *always prefetch*, or *one-block-lookahead prefetch* or *fall-through prefetch*, is evidently an attempt to reduce misses by taking advantage of the spatial locality in sequential accesses. A slight variation on this is to prefetch block $i+1$ only if a miss occurs on block i; this strategy is known as *prefetch-on-miss* and evidently results in fewer prefetches and cache lookups than always-prefetch. With the *tagged-prefetch* strategy, each block has a 1-bit tag that is set to 0 whenever the block is not in the cache, left at 0 when the block is brought into the cache, and set to 1 when the block is referenced; subsequently, prefetching, of the next sequential block, is initiated whenever a tag bit changes from 0 to 1. Tagged-prefetch is there similar to always-prefetch but has the advantages that cache lookups (and repeated prefetching) are avoided for those blocks that have been prefetched and then replaced without being referenced. Other schemes that have been proposed or implemented include *bidirectional prefetch*, in which prefetching is carried out in both the forward and backward directions, and *threaded prefetch*, in which prefetching is based on a list (thread) of addresses, according to the likelihood of subsequent reference: if block i is referenced in a some cycle and block j is referenced in the next cycle, then the address of j is attached to a thread associated with i, and future references to i result in a prefetch of j (and other blocks named in the corresponding thread). For non-sequential accesses, *target prefetch* is a strategy that employs slightly more sophisticated prediction, at a higher cost; a more detailed description of this strategy will be found in Section 3.2.5.

As an example of an implementation of prefetching, we take the two-level cache system in the Power3 microprocessor [30]: Prefetching is here used for both data and instruction caches, i.e. the first-level caches. The instruction cache is two-way interleaved (on line boundaries), and the prefetching attempts to always fetch the line immediately required (in the one bank) and the next line (in the other bank). If the access to the first bank is a hit, but the access to the other bank is not, then a request (on the miss) is made to the unified second-level cache; however, the request is not propagated to the main memory, which action facilitates the cancellation of the request in the event of a branch misprediction. For data accesses, prefetching is for up to eight lines at a time — two for each of four *prefetch streams* — and are from the second-level cache or main memory. A prefetch stream is established whenever a pair of caches misses are found to reference two adjacent blocks of memory. The establishment of a stream consists of entering the corresponding pair of block addresses into a special queue, and prefetching is initiated whenever an address to the data cache matches an address in the queue. A data prefetch may therefore be to the either the block that immediately precedes the current block or to the block that immediately succeeds it; so this is an example of bidirectional-prefetch. The arrangement just described has been shown to improve performance by a factor of up to 33%. Further discussion of the performance of prefetching will be found in section 3.2.5.

3.2.4 Write policies

When a line of the cache is modified, the *write policy* determines when the corresponding memory location is to be modified. There are two main policies: *write-through* (or *store-through*), in which both the main memory and cache are updated at the same time, and *write-back* (or *copy-back*), in which the corresponding memory location is updated only when the block of the cache is selected for replacement. Each of the two policies has its advantages and disadvantages: With a write-back, policy the relative infrequency of block writes to memory means that less memory bandwidth is used, several modifications to the same block may require only one operation to update memory, and the processor is less frequently held up waiting for a STORE operations to be completed. Moreover, not every block that is being replaced need be written back to main memory: only those blocks that have been modified need be written back, and one bit, a *dirty bit*, on each line is sufficient to record this state; this strategy is known as *flagged write-back*, whereas the other is known as *simple write-back*. On the other hand, with a write-though policy the contents of the cache always reflect those of the memory. This means that a consistency problem is easily avoided in those cases where the cache is shared with other processors or input/output devices. Also, with a direct-mapped cache, since the tag and data sections can be accessed concurrently, it may be possible to carry out both tag-check and writing-in at the same time. But only a write-through policy will permit this: if the tag-check indicates a miss, then a write-back policy will result in a loss of the overwritten data, whereas with a write-through policy, there will be a copy in memory of that data.

Another dimension in write policies is the way in which STORE misses are handled: a *write-allocate* policy is one in which a line is allocated for the block to be modified, and a *no-write-allocate* is one that does not do this; a *fetch-*

on-write policy is one in which the block to be modified is fetched into the cache, and a *no-fetch-on-write* policy is one in which this does not happen [14]. (The essential difference between write-allocate and fetch-on-write is that in the former the line is allocated, and the data written in, without old data from the rest also being brought in, whereas in the latter the old data are brought in and then partially overwritten.) The advantage of no-write-allocate is that LOADs of recently written data need not wait for a memory access to be completed, and similarly with no-fetch-on-write there is no delay in waiting for a memory access. Evidently, the best combination is write-allocate/no-fetch-on-write.

3.2.5   *Performance*

We now consider the calculation of the effective access time of a cache/main-memory system. The proportion of accesses that are hits is known as the *hit ratio*, and the proportion that are misses is known as the *miss ratio*. If we let t_h, t_m, and h denote, respectively, the time to process a hit (which is primarily the cache access time), the time to process a miss (which is primarily the main-memory access time), and the hit ratio, then the effective access time is given by the equation

$$t_{eff} = t_h + (1 - h)t_m$$

(t_h is commonly referred to as the *hit time* and t_m as the *miss penalty*.) So to minimise the effective access time, we should maximise the hit ratio (i.e. minimise the miss ratio), minimise the miss penalty, and minimise the hit time.

Table 3.1.   Effect of cache associativity on miss ratio

Cache Size	Degree of Associativity			
(bytes)	1	2	4	8
1024	0.1882	0.1550	0.1432	0.1393
2048	0.1398	0.1169	0.1081	0.1044
4096	0.1033	0.0866	0.0815	0.0786
8192	0.0767	0.0636	0.0591	0.0576
16384	0.0556	0.0457	0.0418	0.0395
32768	0.0407	0.0333	0.0302	0.0289
655361	0.0280	0.0228	0.0211	0.0205
131072	0.0210	0.0144	0.0129	0.0123
262144	0.0142	0.0107	0.0091	0.0085

2-byte block size, LRU replacement

One factor that we should obviously expect to affect the hit ratio is the type of placement policy employed and the degree of associativity: In a direct-mapped cache, the mapping restricts the placement of blocks, so it is possible that conflicts may arise and replacement be required even when the cache is not full. So we may expect a direct-mapped cache to have a smaller hit ratio than a similarly sized cache with a different placement policy. Unlike the case of the direct-mapped cache, replacement in the fully-associative cache will be required only when the cache is full; consequently, we may expect the fully-associative cache to have a better hit ratio than the direct-mapped cache. Replacement may be required in a set-associative cache even if the cache is not full, but this will happen only if the selected set is full. We should therefore expect that a set-associative cache will have a hit ratio somewhere between those of similarly-sized direct-mapped and fully-associative caches and that the hit ratio will be higher for higher degrees of associativity. Table 3.1 gives some data (from simulations) that confirms all this [10].

Although miss ratio diminishes with associativity, direct-mapped caches, because of their simplicity and the fact that data-readout and tag-checking can be carried out in parallel, can have smaller hit times than similarly sized set-associative and fully-associative caches, depending on the realization technology; the absence of many comparators also means that a direct-mapped cache will take up relative less area [11]. It is for these reasons that several manufacturers after implementing only set-associative caches in a number of machines now also use direct-mapped caches. Examples of such implementations are in the DEC Alpha (Section 3.2.6.3), the Sun Ultra-SPARCs (Section 3.2.6.4), as well as a number of other current high-performance machines.

A second way to minimise the miss ratio low is to maximise the exploitation of spatial locality. This may be done by either increasing the block size or by prefetching (as indicated above). The former is useful only up to a certain point: for a given cache size, increasing the block size means a reduction in the number of cache lines and, therefore, a reduction in the exploitation of temporal locality; at some point the loss from the failure to exploit temporal locality outweighs the gain from spatial locality, and the miss ratio increases. A large block size also requires a larger transfer time and can therefore increase the miss penalty. The simulation data in Table 3.2. shows the effect of line size on miss ratio; the data given are for ratios of miss ratios: for each cache size and block size, the figure given is the ratio of the miss ratio for that cache size and block size and the miss ratio for the same cache size and half the block size [28]. (Note that Tables 3.1 and 3.2 are from different sources.)

A third technique that has been shown to be effective in reducing the miss ratio when the write policy is write-through is the use of a *write cache*, which is a small fully-associative cache (with LRU replacement) placed immediately behind the main cache (Figure 3.13). In the case of a direct-mapped cache, the similar placement of a *victim cache*, which is a small fully-associative cache for lines being replaced in a write-back policy, has been shown to reduce the miss ratio (Figure 3.13). The victim or write cache is searched at the same time as the main cache, and the request is processed at the next level in the memory hierarchy only if a miss occurs in both caches [14,15].

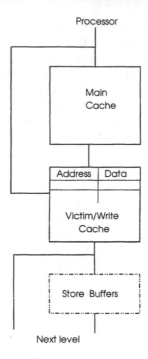

Figure 3.13.   Placement of write or victim cache

Table 3.2.   Effect of block size on miss ratio

Cache Size	Block Size				
(bytes)	8	16	32	64	128
128	0.66	0.78	0.94	1.24	2.15
256	0.64	0.73	0.86	1.04	1.44
512	0.61	0.70	0.80	0.93	1.17
1024	0.60	0.67	0.76	0.86	1.02
2048	0.58	0.65	0.72	0.81	0.92
8192	0.56	0.62	0.68	0.74	0.82
16384	0.55	0.61	0.66	0.72	0.78
32768	0.54	0.60	0.65	0.70	0.75

fully-associative cache, LRU replacement

As indicated above, prefetching can also be effective in reducing miss ratio. Tables 3.3 and 3.4 give some results from a study that compare the various prefetching strategies in split caches [31]. The benchmarks used are a suites of computer-aided design programs (CAD), compiler tools (COMP), floating-point applications (FP), text processing programs (TEXT), and Unix utilities (UNIX). Table 3.3. is based on a "baseline system" in which the parameters are in line with current technolgies and design; and Table 3.4. is based on an idealized system, in which every parameter has been chosen to give the best results. Table 3.3 shows that, as might be expected, prefetching is most profitable with program execution where there are many sequential accesses — the FP suite consists of programs that mainly manipulate vectors and matrices — and less so for programs (e.g. those in the UNIX suite) that do not. Also, bidirectional prefetching, because it tends to favour non-sequentiality, works better than other strategies. In Table 3.4 the results have improved for all the strategies, and in particular for always-prefetch and tagged-prefetch. Together the two tables show that when there are few resources, conservative strategies (that issue relative few prefetch requests) perform well, but the aggressive strategies suffer from too many conflicts; on the other hand, when there is no shortage of resources, the aggressive strategies perform quite well.

Table 3.3.   Performance of prefetching in baseline system

Cache Prefetching Strategy	Relative MCPI					
	Trace					Trace Avg.
	cad	*comp*	*fp*	*text*	*unix*	
none	1.0000	1.0000	1.0000	1.0000	1.0000	1.0000
always	1.5914	1.2924	0.8821	1.1770	1.7009	1.2943
miss	1.2686	1.1505	0.9523	1.1306	1.6159	1.2049
tag	1.3774	1.1561	0.8793	1.1495	1.6695	1.2186
bi-dir	1.4329	1.1979	0.8885	1.1712	1.0818	1.1408
thread	1.0004	0.9968	0.9979	0.9933	1.3515	1.10596
Avg.	1.3186	1.1547	0.9188	1.1222	1.4633	1.1836
	Absolute MCPI					
none	0.0863	0.0983	0.1759	0.0306	0.1460	0.1704
always	0.1373	0.1271	0.1552	0.0360	0.2484	0.1408
miss	0.1094	0.1131	0.1675	0.0346	0.2360	0.1321
tag	0.1188	0.1137	0.1547	0.0352	0.2438	0.1332
bi-dir	0.1236	0.1178	0.1563	0.0358	0.1580	0.1183
thread	0.0863	0.0980	0.1755	0.0304	0.1974	0.1175
Avg.	0.1151	0.1139	0.1618	0.0344	0.2167	0.1284

MCPI = (total memory-access penalty – processor stalls from cache-write hits)/(instructions executed); Relative MCPI = (MCPI with prefetching)/(MCPI without prefetching)
cache: 64K-byte, 64-byte line, 8-way associative, 2-port tag array, 1-port data array
memory: 4 banks, 16-cycle (cpu) miss-penalty
prefetch distance: 1 block

Another study of prefetching in instruction caches has compared always-prefetch and target-prefetch in an instruction cache in detail and concluded the following [12]: For fall-through prefetching, miss rates are reduced significantly — by about one-third, on average — when prefetching is used; the exploitation of spatial locality through the prefetching means that large line sizes (that would have to be otherwise used to achieve the same effect when

Table 3.4. Performance of prefetching in idealized system

Cache	Relative MCPI					
Prefetching	Trace					Trace
Strategy	cad	comp	fp	text	unix	Avg.
none	1.0000	1.0000	1.0000	1.0000	1.0000	1.0000
always	0.5941	0.5133	0.3110	0.4295	0.8686	0.5433
miss	0.6980	0.6313	0.5202	0.6031	0.9186	0.6743
tag	0.5902	0.5121	0.3111	0.4428	0.8678	0.5448
bi-dir	0.8069	0.7308	0.4583	0.6879	0.8912	0.7150
thread	0.7836	0.7076	0.7882	0.7073	0.8941	0.7761
Avg.	0.6946	0.6190	0.4778	0.5741	0.8881	0.6507
	Absolute MCPI					
none	0.4292	0.6273	2.3140	0.2090	0.4362	0.8031
always	0.2550	0.3220	0.7197	0.0897	0.3789	0.3531
miss	0.2996	0.3960	1.2038	0.1260	0.4007	0.4852
tag	0.2533	0.3212	0.7200	0.0925	0.3786	0.3531
bi-dir	0.3463	0.4584	1.0605	0.1437	0.3887	0.4796
thread	0.3363	0.4438	1.8238	0.1478	0.3900	0.6284
Avg.	0.2981	0.3883	1.1056	0.1200	0.3874	0.4599

cache: 1M-byte, 16-byte line, 16-way associative, 2-port tag array, 2-port data array
memory: 16 banks, 64-cycle (cpu) miss-penalty
prefetch distance: 1 block

there is no prefetching) offer relatively little advantage; for caches with the best performance, the relatively smaller line sizes when prefetching is used means that memory traffic is less than with no prefetching (and larger line sizes); and high efficiencies are achieved with low miss rates. For target prefetching, the target-prediction table used in the study has the organization in which each line consists of a ⟨current-line-address, next-line-address⟩ pair, and the table is updated whenever a line change occurs in the program-counter's value. When a line change occurs, the address of that line is used to access the table, via the current-address line. (The placement policy for the table may be fully-associative, set-associative, or direct-mapped, irrespective of the cache's placement policy; and in the case of the direct-mapped cache, the tags need not be stored in the table, which means that every access is a hit.) If there is a corresponding entry, then the target-line address there is used for prefetching; otherwise no prefetching takes place. In the case that a cache miss coincides with a valid prefetch request, both missed line and predicted prefetch line are brought from memory. Tables 3.5 and 3.6 summarizes the results obtained with fully-associative (random replacement) table and direct-mapped tables, both for fully-associative and direct-mapped caches. Compared with fall-through prefetching, the results show that for target-prefetch the overall performance is better when the prediction table is fully-associative, that optimal line sizes tend to be larger, and that efficiencies are higher for small caches. The study also shows that further refinement that includes a combination of both prefetch strategies to prefetch along both paths gives better performance that either strategy used alone.

Table 3.5.   Miss rates with target prefetching

Cache		Fully Associative Caches								
and Line		Fully Assoc. Table			Dir. Map. Table/Tags			Dir. Map. Table		
Sizes		8	32	128	8	32	128	8	32	128
128	16	2.14	1.36	0.96	2.26	1.58	1.12	2.33	1.72	1.18
	32	1.49	1.13	0.92	1.66	1.18	0.99	1.78	1.26	1.05
	64	1.35	1.21	1.13	1.39	1.25	1.17	1.47	1.27	1.17
256	16	1.65	1.04	0.62	1.70	1.23	0.77	1.71	1.27	0.82
	32	1.08	0.76	0.55	1.17	0.81	0.62	1.23	0.91	0.66
	64	0.81	0.72	0.63	1.85	0.76	0.66	0.92	0.78	0.66
512	16	0.86	0.72	0.33	0.86	0.75	0.48	0.86	0.78	0.49
	32	0.65	0.45	0.25	0.65	0.50	0.31	0.69	0.56	0.34
	64	0.47	0.36	0.28	0.51	0.40	0.31	0.58	0.46	0.35
1024	16	0.58	0.56	0.26	0.58	0.57	0.39	0.58	0.66	0.37
	32	0.41	0.34	0.14	0.42	0.36	0.20	0.42	0.36	0.21
	64	0.30	0.20	0.12	0.30	0.24	0.15	0.31	0.24	0.16
2048	16	0.28	0.28	0.16	0.28	0.28	0.21	0.28	0.28	0.26
	32	0.29	0.26	0.11	0.29	0.27	0.15	0.29	0.27	0.15
	64	0.20	0.13	0.07	0.20	0.16	0.09	0.20	0.17	0.10
		Direct Mapped Caches								
128	16	2.16	1.42	0.99	2.28	1.64	1.14	2.41	1.70	1.16
	32	1.65	1.32	1.12	1.80	1.38	1.19	1.87	1.41	1.19
	64	1.50	1.37	1.30	1.55	1.41	1.33	1.63	1.42	1.34
256	16	1.66	1.13	0.70	1.79	1.35	0.87	1.88	1.38	0.87
	32	1.20	0.94	0.73	1.35	0.99	0.81	1.42	1.02	0.80
	64	0.00	0.89	0.81	1.05	0.93	0.84	1.12	0.94	0.86
512	16	0.05	0.80	0.45	1.14	1.04	0.60	1.31	0.09	0.62
	32	0.78	0.62	0.41	0.85	0.67	0.48	1.20	0.72	0.49
	64	0.65	0.54	0.66	0.69	0.58	0.49	0.74	0.62	0.51
1024	16	0.66	0.55	0.28	0.67	0.63	0.43	0.95	0.67	0.44
	32	0.46	0.37	0.18	0.47	0.42	0.25	0.96	0.45	0.27
	64	0.35	0.26	0.18	0.36	0.29	0.21	0.77	0.32	0.22
2048	16	0.41	0.36	0.18	0.42	0.39	0.33	0.63	0.41	0.35
	32	0.29	0.23	0.10	0.30	0.26	0.16	0.40	0.28	0.18
	64	0.22	0.14	0.09	0.23	0.19	0.11	0.27	0.22	0.44

miss rate = instruction-misses/instruction-fetches

There are a number of other ways in which the effect of the miss penalty can be minimized. One that is increasingly being used in current machines is *multi-level caches*, in which the miss penalty is reduced by the placement of one or more levels of cache between the processor and the main memory. The extra levels have capacities and access times that increase and decrease, respectively, as one gets farther from the processor and therefore capture items that would otherwise be in main memory; a given level is therefore accessed only when there are misses at all preceding levels. The majority of current implementations of multi-level caches have just two levels, with an optional third level in a few). For a two-level cache, the performance equation above is now replaced with

$$t_{eff} = t_{h1} + (1 - h_1)[t_{h2} + (1 - h_2)t_m]$$

where t_{h1} and t_{h2} are the hit times for the first and second levels, respectively, h_1 and h_2 are the corresponding hit ratios, and t_m is the miss penalty at the second level. (The extension to three or more levels is along similar lines.) The current trend in the design of several high-performance processors is to have a direct-mapped cache at the first-level and a set-associative cache at the second level.

Another way of dealing with the miss penalty is to try and mask its effect, instead of reducing the basic latency. An increasing number of machines do this by having a a *non-blocking* (or *lock-up free*) caches, in which the miss penalty is masked by arranging it so that the cache can continue to service other requests while a miss is being processed [17]. A non-blocking cache is, however, generally more complex than a blocking one, as it must buffer responses and requests and also maintain a consistency-ordering between them: in general, the order of hits and misses will not correspond to the ordering of requests from the processor. The most direct implementation that easily allows this is a structure that permits concurrent search between addresses of new requests and addresses corresponding to requests that have been held up.

The miss penalty can also be masked by pipelining the cache or the cache-memory interface. A straightforward way of achieving the latter is to place buffers between the cache and the main memory: a first-in/first-out *load buffer* (or *read buffer*) to hold addresses of LOADs waiting to be processed and a first-in/first-out *store buffer* (or *write buffer*) to hold addresses and data of waiting STOREs. (A non-blocking cache will, of course, require similar type of buffers.) Where such buffering is employed, it may be used to implement optimizations in processing LOAD/STORE sequences, in manner similar to those described above for an interleaved memory system; this is currently done in a number of high-performance machines, an example being the Motorola 88110 (Section 3.2.5.3). Even if such buffering is not used to the full extent possible, it may still be useful in reducing traffic to memory (or next cache level): a *coalescing buffer* is one that collects a number of STOREs to the same block before writing back to the next level of the memory hierarchy.

Lastly, the hit time may be minimised by employing one or more of the following techniques: implementing a small and simple cache, pipelining the cache, interleaving the cache, and accessing the cache with virtual addresses (instead of real addresses) and therefore eliminating the address-translation time for all hits. The first three of these techniques are implemented in many machines, but the last is usually problematic: a *synonym* (or *aliasing*) problem can occur if two different virtual addresses that refer to the same physical address are in the cache at the same time; in such a case, STORES to a single address can cause inconsistency, unless special care is taken. Nevertheless, a number of high-performance machines do implement virtually-addressed caches of one sort or another. The MIPS R4000 direct-mapped data cache exemplifies the type of partial virtual-addressing that is used in a number of machines. Here the virtual address is used only to index the cache, with the tag being taken from the real address, and the performance penalty of address translation (which is required before the tag-check) is minimized[8] by carrying it out in parallel with the data-readout. To avoid inconsistency, the second-level cache in the machine is real-addressed and primary cache is constrained so that it cannot have more than one copy of a line at the second level. On the other hand, the Silicon Graphics TFP has an instruction cache that is both virtually-indexed and virtually-tagged; so the address-translation penalty is eliminated for all hits in the cache. The aliasing problem is not a major problem in this machine: STORES into an instruction stream are rare, and the underlying architecture requires that they be carried out only under operating-system control.

3.2.6   Case studies

In this section, we look at four examples of practical cache implementations in pipelined machines. The first is the Manchester MU5 Name Store and is an example of a fully-associative cache [22]. The second is the cache system of the DEC Alpha 21164 and is is an example of a two-level cache that combines direct-mapping and set-associative placement and which also uses a sort of victim cache at the second-level cache [8]. The third example is the cache system of the the Motorola 88110; this is a cache with limited non-blocking, optimization of LOAD/STORE sequences, and program-control of the cache [7]. The last example is the cache system of the Sun UltraSPARC-1; this has a two-level cache (although, unlike the Alpha, the second level is off-chip) and Load/Store buffering with various optimizations [33].

3.2.6.1   The MU5 Name Store.   The MU5 is a highly pipelined machine designed on the basis of a careful analysis of high-level language structures and program properties; a brief description of the machine is given in Chapter 1. Studies during the design of the MU5 computer indicated that, over a large range of programs, about 80% of operand accesses were to named operands (scalar variables and descriptors of structured data, such as vectors) and that in a typical program only a small number of names were in use at any given time: 95% of the named references were to fewer than 35% of the names used.

---

[8]The penalty cannot be eliminated in all cases, as there will be times when a miss occurs in the Translation Lookaside Buffer.

Accordingly, the decision was made to buffer these in a fully-associative store that would hold scalars and array descriptors. In the implementation, there are two *name stores*, one for the *primary pipeline*, which is used to process named operands, and one for the *secondary pipeline*, which is used to process operands accessed relative to a descriptor. In what follows, we shall consider only the design of the Name Store in the primary pipeline.

The Name Store makes up two stages of the primary pipeline and is accessed with 20-bit virtual addresses: 4 bits of the address identify the process concerned, 15 bits specify a word within a virtual-memory segment, and 1 bit is used to distinguish between interrupt procedures and other code; the interrupt procedures use four reserved lines of the store. The store consists of 32 lines; the number was determined by technological constraints and studies at the time of design that showed that this number was sufficient to get a very high hit rate. The organization of the Name Store is shown in Figure 3.14. Each line of the store has three associated bits: the *line used* bit, which is held in the LU register, indicates whether or not the line holds valid data; the *line altered* bit, which is held in the LA register, indicates whether or not the contents of the line have been modified; and the *line pointer* bit, which is held in the LP register, indicates whether or not the line is the next candidate for replacement when this becomes necessary. The Value field holds a 64-bit data word or descriptor.

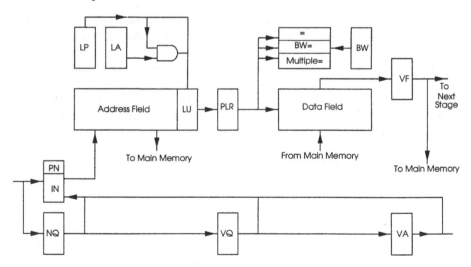

Figure 3.14.   Name Store in MU5 Primary Pipeline

During normal operation, an address held in the PN-IN register is presented for association to the Address field of the store. If there is a match and the corresponding LU-bit is set, then in the next pipeline cycle a bit in the register Prop Line Register (PLR) is set to identify the line on which the match occurred, and in the following cycle the contents of the selected line are then read into the Value Field (VF) register. The write policy used is write-back;

and the replacement algorithm used is NRU; more sophisticated algorithms, such as LRU, were found to yield little performance improvements while being considerably more costly.

If association fails or the LU-bit for the selected line is not set, then the required datum is fetched from memory and written into the line identified by the LP register. The details of this operation are as follows. (At this time the address of the required operand will be in the VA register.) First, the stages of the pipeline before the Name Store are stalled. Next, the contents of the VA register are copied back into the PN-IN register and those of the LP register are copied into PLR. If the line identified by PLR at this point is one that is the target of an outstanding WRITE, then the next line is selected instead; this is done by copying the contents of PLR into LP, with a 1-bit shift, and then copying the contents of LP back into PLR. A check is then made of the LA register to determine whether or not the contents of the line have been altered; if they have, then the contents of the Address and Value fields are sent to main memory, which then updates the corresponding memory location. (At this point the selected line is ready for writing.) When the required data arrives from memory, it is written into the selected line, the corresponding bits in LU and LA are set to 1 and 0, respectively, and the contents of PLR are copied into LP, with a 1-bit shift. At the completion of the these steps, the contents of PLR and PN-IN no longer correspond to the position of the instructions in those stages of the pipeline and must be restored. The restoration is done by copying the address in the VQ register into PN-IN, allowing for association, re-strobing PLR, and then copying the contents of NQ into PN-IN. The stalled stages of the pipeline are then re-started, and normal sequencing of the pipeline is resumed.

The description above only covers the case where the operand comes from memory. In some cases the operand will be in the Name Store in the secondary pipeline, and this requires a different set of actions; for brevity, we here omit discussion of these.

3.2.6.2   The DEC Alpha caches.   The DEC Alpha 21164 is a recent implementation of DEC's Alpha architecture. A brief description of its overall organization is given in Section 1.5.3. The data cache, which is what we discuss in the following, consists of two levels (with an optional third level) that make up parts of several stages (Stages 4 through 8) of the processing pipeline, whereas the third level is off-chip. The primary cache system consists of an 8-Kbyte, direct-mapped, write-through cache, a structure (the Miss Address File) that handles misses, and a 6-block write buffer (Figure 3.15). The cache has a line width is 32 bytes. The tag field of the cache memory holds virtual addresses, which are are translated into physical addresses in the Data Translation Buffer.

Virtual address translation starts early in the Stage 4 of the pipeline, and cache access begins later in the same stage and ends in the next stage. If a hit occurs, then in the next stage data are transferred into the register file (for a

LOAD) or into the cache (for a STORE). Misses are handled by the Miss Address File in conjunction with the Cache-Control/Bus-Interface Unit (Figure 3.16).

The Miss Address File is essentially a Load Buffer with two main sections that serve as a buffer into the Cache-Control/Bus-Interface Unit: one section consists of a 6-slot store that holds information on LOAD misses, and the other is a 4-slot store that holds addresses for instruction fetches. An entry for a LOAD miss consists of the physical memory address, the destination-register number, and the instruction type; the last of these is used to assist in properly formatting the returned data, selecting the appropriate register file, and updating the information in the instruction-issue unit. The entry for an instruction-fetch miss consists of just a real address. LOAD-miss entries are deleted only when the corresponding data has been returned, whereas instruction-fetch entries are deleted as soon as the request has been accepted by the Cache-Control/Bus-Interface Unit.

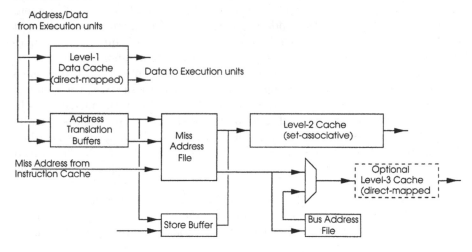

Figure 3.15.   DEC Alpha 21164 cache system

In order to maximise performance, the Miss Address File can combine several requests to the Cache-Control/Bus-Interface Unit into a single one: a single LOAD can request no more than eight bytes, so up to four LOAD-misses can be combined into one request if they reference the same 32-byte block. This coalescing is realized in logic that detects addresses to blocks for which there are already waiting requests and then adds new destination addresses to the list of waiting ones without making any new queue entries; the logic can combine two such requests in each cycle.

The Cache-Control/Bus-Interface Unit contains the second-level cache and a number of other units. The second-level cache has a capacity of 96 Kbytes, in 64-byte blocks, and is 3-way set-associative; the replacement policy is Random; and the write policy is write-back. The other units in Figure 3.16 are a Cache Arbiter Unit, which arbitrates between requests to the cache; the Bus Interface Unit Sequencer, which controls the movement of data to and from the

processor; the Victim Address File, which holds entries (*victims*) rejected from
the second-level cache, information for memory broadcast STOREs, and external
cache-coherence commands that need data from the second-level cache; the Bus
Interface Unit Address File, which queues misses from the second-level cache;
and the Write Buffer Unit, which handles second-level STOREs and assists with
the maintenance of coherency.

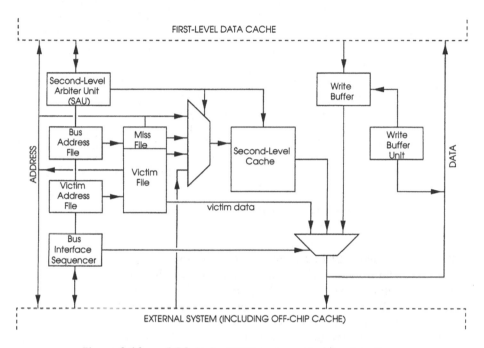

Figure 3.16.   DEC Alpha 21164 cache-control/bus-interface

The operation of the second-level cache is exemplified by the sequence for
a LOAD operation: First, the Cache Arbiter Unit operates in Stage 5 of the
pipeline to select the source for the next address. This is then followed by the
reading (in Stage 6) of the cache's tag field and (in Stage 7) the determination
of whether or not there was a hit. In the event of a hit, the first half of the
block is read out (in Stage 8) and placed on the bus (in Stage 9), with the
second half following in the next pipeline cycle, and the cache-fill is carried out
in Stages 10 and 11. For a STORE access that results in a hit, the processing
sequence consists of arbitration in Stage 5, hit-determination in Stage 6, data
onto the bus in Stage 7, and writing (in two cycles) in Stages 8 and 9.

In the DEC Alpha 21264 (the next implementation of the Alpha architecture), the data cache has two ports that allow it to process any combination
of two LOADs and STOREs in each cycle; pipelining allows the cache to start a
new access in each half-cycle. To support the two ports, the address translation
buffer and the cache-tag array have been replicated. Cache misses are held in

an 8-entry buffer, and rejected blocks are held in an 8-entry victim buffer. Both
first-level and second-level caches are non-blocking.

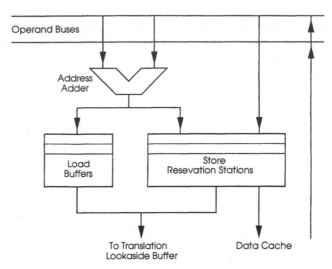

Figure 3.17.   Load/store unit in Motorola 88110

3.2.6.3 Motorola 88110 caches.

The Motorola 88110 has an 8Kbyte in-
struction cache and an 8Kbyte data cache. The addressing in both caches is
based on virtual-indexing and real-tagging; so address translation can proceed
in parallel with the retrieval of the tags and data, with final line selection being made after the tag-check. The processor accesses the data cache via the
Load/Store unit, whose organization is shown in Figure 3.17. The unit consists of an address-adder, a 4-entry Load Buffer for LOAD operations, and three
Reservation Stations[9] that are used for STORE operations; the main purpose
of the last two components are to provide pipelined buffering to hide the relatively long latency of memory operations. The Load Buffer is managed as a
first-in/first-out queue, and each entry holds the virtual address of a memory
location from which data are to be loaded into a register. The Reservation
Stations are also managed as a first-in/first-out queue, and each holds data
and the virtual address of the memory location into which the data is to be
written; the address part of an entry may be made before the data is available,
but no operation is carried out until both data and address are available.

The operation of the two queues with respect to one another is not synchro-
nized, in that LOADs may proceed ahead STOREs that are stalled in the Reservation Stations while awaiting the arrival of data from the execution units.
Evidently, this can be done only in cases where there is no dependence between

---

[9]Reservation stations, in general, are discussed in detail in Chapter 5. It will suffice here to
simply view them as a STORE buffer.

data to be loaded and data to be stored. Accordingly, the system includes a comparator that compares the address of each LOAD with all the addresses in the Reservation Stations and prevents the LOAD from being carried out if there is a match.

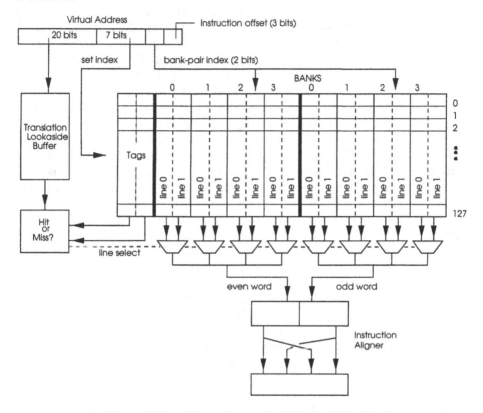

Figure 3.18. Instruction cache in Motorola 88110

The instruction cache is 2-way set-associative, with a total of 128 sets and a line size of 32 bytes (8 instructions). The cache can provide a pair of instructions (two 32-bit words) to the pipeline in each cycle, except on rare instances when the pair straddles a cache-line boundary. To minimize latency on a cache miss, a burst transaction is used on the bus. During this transaction, two instructions are transferred on each cycle, beginning with the pair containing the missed instruction and continuing (with possible wraparound) until the cache line is filled; as the instructions are delivered from memory, they are also forwarded to the pipeline, so that there is no break in the instruction flow. The arrangement by which one instruction-pair per cycle is delivered to the pipeline is as follows (Figure 3.18). The cache is divided into 8 banks, each consisting of 128 64-bit rows. Four of the banks hold the even-numbered words, and the other four hold the the odd-numbered words; so to return an instruction-pair, the "odd" banks return one instruction, and the "even" banks return the other instruction. The

Aligner at the output of the cache serves two purposes: The primary one is to swap, whenever necessary, the instructions read out out so that they enter the pipeline in the correct order, irrespective of their alignment — for example when instruction $i + 1$ comes from an "even" bank and instruction i comes the "odd bank", instead of the other way around. The other purpose is to permit, by swapping bytes if necessary, the processor to be used in environments in which addressing is little-endian rather the big-endian of Motorola's architecture.

The organization of the data cache is similar to that of the instruction cache. The write policy used is write-back/write-allocate, but the program has the option of selectively changing this (by setting a bit in an instructions opcode) to write-through. In the write-through mode, if there is a hit in the cache, then both cache and memory are updated; but if there is a miss, then only the memory is updated, the idea being to be able to avoid having in the cache data that it is known will not be used again. The replacement policy is essentially random, although it attempts to first select invalid lines. The cache is lock-up free in a limited way: if a LOAD bypasses and STORE and misses in the cache, the cache is decoupled from the memory bus so that the STORE can still access memory while the LOAD is stalled, and the same applies to a STORE miss followed by a LOAD hit. The machine also has a facility, the TOUCH-LOAD instruction, that allows the user to prefetch data data into the cache, thus avoiding some stalls that would occur if a demand-fetch policy were to be used at all times.

The Load/Store unit also implements a number of instructions that allow program-controlled allocation and deallocation of lines in the data cache. For example, an ALLOCATE instruction may be used to allocate a line of cache without a corresponding fetch from memory, and a LINE-FLUSH instruction overrides the normal replacement algorithm in the ejection of blocks from the cache.

3.2.6.4   UltraSPARC-1 data cache.   The UltraSPARC-1 is designed for a two-level cache, but only the first-level cache is on-chip. The primary data cache is non-blocking and direct-mapped with a write-through/no-write-allocate policy. The cache has total size of 16 Kbytes, divided into 32-byte lines, each of two 16-byte sub-blocks. The cache is virtually-indexed and real-tagged, with the task of dealing with aliasing left to the operating system. The write policy is write-through, which was chosen to simplify the design. The organization of the cache system is shown in Figure 3.19. The Load Buffer is a 9-entry first-in/first-out queue that is used to support non-blocking LOADs. Each entry of the buffer consists of an address from which data is to be read but for which there is no entry in the cache. LOAD operations are completed out-of-order, with respect to instructions that do not depend on their results, but not with respect to other LOADs; allowing the latter would have affected both the complexity and cycle time of the system. The Store Buffer is an 8-line first-in/first-out queue that holds data-address pairs for STORE operations until an update can be carried out in the primary or secondary cache. The management of the buffer includes a feature that allows for the combination of any two or more entries in the Store Buffer into a single entry if they refer to the same 16-byte block. Since the primary cache has a write-through policy, each STORE must have its effect reflected also in the secondary cache; thus there are two steps for a STORE: checking the secondary cache for a hit/miss and updating that cache with the new data.

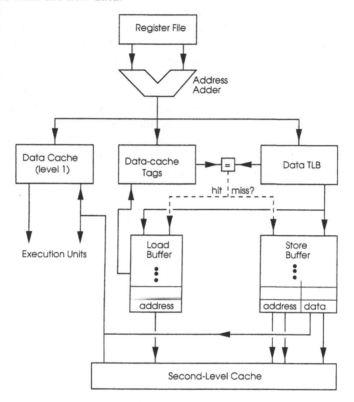

Figure 3.19.   Cache system in the UltraSPARC-1

The secondary cache is direct-mapped. The addressing is real-indexing with real-tagging; and the size is at least 0.5 Mbytes and at most 4 Mbytes, arranged in 64-byte lines. The write policy is write-back/write-allocate. The cache is organized as a five-stage pipeline: Cache Request, Address Transfer, Access Data SRAM, Data Transfer, Tag Check. The Tag and the Data parts of the cache are arranged in such a way that a tag-check can occur in parallel with data-writing from an earlier operation; being able to start the tag check as soon as an address has been translated allows a throughput of one STORE per cycle.

3.3 SUMMARY

In this chapter we have discussed the two main techniques used in the design high-performance memory systems: interleaving of main memory and the use of caches. Although interleaving was initially developed for main memory, as performance requirements increase, similar (and related) techniques are increasingly being used with caches. Interleaved main memory will be further discussed, with respect to data placement, in vector-pipelined machines (Chapter 6).

Of the different types of cache systems, fully-associative caches are rarely used for instructions or data, as their complexity increase with size while the performance decreases. Nevertheless there have been a few implementations, even in recent machines: for example the Hitachi Gmicro/100 has a 1024-line fully-associative instruction cache and a 32-line fully-associative stack-data cache. Fully-associative caches are, however, much used for special applications, such as address translation and branch prediction. Until recently the most common type of cache was the single-level set-associative. But as the complexity and performance of such caches have become inadequate with the large sizes used in current machines, the current trend is to implement multi-level caches, with direct-mapping at the level closest to the processor. Other trends that may be expected are the increased implementation of non-blocking caches and high pipelining of the first-level cache.

4 CONTROL FLOW: BRANCHING AND CONTROL HAZARDS

For an instruction pipeline to attain its maximum performance, it is, at the very least, necessary that it be supplied with instructions at a rate that matches its maximum processing rate. The main impediment to ensuring adequate instruction-supply is usually the high access time (relative to the pipeline cycle time) of the memory from which instructions are fetched. At any given moment, the addresses of the next instructions required are easy to determine if there are no branch (control-transfer)[1] instructions involved: simply incrementing the program counter, or similar addressing register, suffices. A branch instruction, on the other hand, presents a problem, since the addresses of the following instructions cannot be known with absolute certainty until after the branch has been executed; furthermore, the execution may depend on a condition yet to be determined by preceding instructions. Consequently, unless special measures are taken, a branch instruction will introduce a gap — the delay of which we shall term the *branch latency* — in the flow of instructions. In this chapter we shall discuss a number of measures for dealing with this, which is arguably the hardest problem in the design of high-performance instruction pipelines [116].

Without the use of special techniques, there are two basic strategies for dealing with problem of branches: The first is to continue processing instructions

---

[1]Sometimes *branch* is used for control-transfers relative to the program counter, and *jump* is used for absolute transfers. Unless specified, we shall use the *branch* to refer to any type of control-transfer.

in the straight-ahead path and flush the pipeline (i.e. *squash* or *nullify* the partially processed instructions) if the branch should have been taken. Alternatively, we may assume that the branch will be taken, fetch the instructions from the target address, and then squash them if necessary; however, assuming "not-taken" is simpler, in that it does not require the computation of the target address. The second strategy is to simply stop processing new instructions until the branch-target address has been determined. This is not a very good strategy, as stalling the pipeline can have a serious effect on performance: Suppose, for example, that the delay involved in initiating a new instruction stream is 100ns in a pipeline with a cycle time of 10ns and the probability that an instruction causes a transfer of control is 0.3. Then the effective cycle time will be $0.3 \times 100 + 0.7 \times 10 = 37$ns — a performance drop by a factor of about four. "Proceed-and-squash" also has similar implications for performance, since there will be times when the alternate path is the correct one; and there are also cost implications, because some means of saving and restoring machine state may be required. Nevertheless, "proceed-and-squash" is likely to be better than simply stalling the pipeline, especially, if the branch latency is high, as there will be times when the straight-ahead path is the correct one.

If we modify the simple analysis of Section 1.2 to an n-stage pipeline with a cycle time of τ_c and an instruction-stream initiation time of τ_m, then the total time taken to process M instructions is

$$T = \tau_m + n\tau_c + (M - 1)\tau_c$$

This assumes the absence of branch instructions. Suppose branch instructions occur with a probability of p. Then, in the absence of any special techniques to deal with them — that is, with stalling — and assuming that all branches are taken, the total time taken may be approximated by

$$T' = (pM + 1)\tau_m + (pM + 1)n\tau_c + [M(1 - p) - 1]\tau_c$$

Here, the first term is the time required to fetch from memory the first instruction and each of the branch targets, the second term is the start-up time for each of these instructions, and the last term is the processing time for the remaining instructions. A consideration of typical values of p (see Table 4.3) and τ_m (especially the component due to memory access time) shows that the dominant terms in this equation are the first and the second, and, therefore, the previous analysis of speed-up does not hold. The objective in designing a good instruction-supply unit is to mask out the effects of the terms in p so that T' approaches T and the earlier analysis of performance therefore holds. This is especially important, as pipelines get deeper and wider, and, therefore, the probability that an instruction in progress is a branch also increases: The cost of a branch instruction is not just the cycles lost; rather, it is the number of *instruction slots* lost (i.e. the number of instructions that could have been issued in the time lost). If a machine is capable of issuing m instructions in each cycle and loses k cycles on a branch, then the cost is $k \times m$ instruction slots; as an example, in the PowerPC 604, k is 3 and m is 4, so the cost is 12 instruction slots. In other words, the cost of a branch increases with both the length of the pipeline and the degree to which it is superscalar.

From the above, we may say that the delay introduced by a branch instruction has three main components. The first is the *address latency*, which is the time from when the the branch instruction is decoded to the time when its target address is determined, and which may include the time — the *condition latency* — required to evaluate some condition. The second component is the *fetch latency*, which is the time from when the target address is determined to when the target instructions enter the pipeline. And the third component is the *pipeline latency*, which is the delay introduced by gaps in the pipeline when the control-transfer takes place (i.e. the time required to refill the pipeline). The following important points should be noted about these definitions: First, the address latency may be negative, which means that the target address is computed before the corresponding branch instructions is decoded; therefore, the fetch latency does not necessarily include or overlap with the address latency. Second, if the address latency is sufficiently negative, then the target instructions can be fetched early (and stored near the pipeline), thus reducing, or perhaps even completely eliminating, the fetch latency. Third, if the fetch latency is sufficiently reduced, then the pipeline latency can also be reduced or eliminated. Lastly, if the execution of branch instructions is overlapped with the execution of other instructions, then all three latencies can, in essence, be reduced; such overlapped execution is common in current superscalar machines.

In this chapter we shall discuss ways of reducing the effects of the three latencies above. The first section of the chapter deals with the effect of pipeline length and the position of the *control point* (the point at which the branch instructions executed are executed)[2] on the pipeline latency. The second section discusses the use of hardware buffers to reduce fetch latency. The third section deals with software-supported techniques that that involve the re-organization of code in such a way as to mask the fetch and address latencies and, by so doing, also mask the pipeline latency. The fourth section covers the use of target-address prediction to reduce the address latencies in such a way that both fetch and pipeline latencies are also reduced. The fifth section deals with the removal of branches from the instruction stream, the sixth section is a brief discussion of other solutions to the branching problem, and the seventh is a discussion of how the instruction-set architecture can affect whatever solution is adopted. The last section is a summary.

---

[2]In many machines this is also the point at which the program counter is updated.

4.1 PIPELINE LENGTH AND LOCATION OF CONTROL POINT

In this section, we shall discuss the effects of the pipeline length and the location of the control point on the pipeline latency. There are three main factors that determine the effect of branches. The first two, as the equation above for T' shows, are the length of the pipeline and the instruction-access time (e.g. from cache or memory); the third is the position of the control point [56].

If the program counter is also altered at the control point, then once an instruction has proceeded beyond that stage, it is considered to have been executed, which means that, ideally, no instruction whose effects cannot be undone should be allowed to proceed beyond that point. The pipeline latency for a branch instruction is therefore closely related to the start-up time and the flush time, but as we shall see below, it is not always just the one or always the other or the sum of both.

When a conditional branch instruction is decoded, either no further instructions can be allowed into the pipeline or if any are, then they must be discarded (and their effects undone) if they are not from the correct path. In either case a start-up latency is incurred when the instructions at the branch target-address are eventually brought into the pipeline, and this latency increases linearly with the length of the pipeline; it is partly for this reason that some machines without sophisticated branch-handling techniques have been designed with fairly short pipelines. The effect of the drain time, which effect is also influenced by the position of the control point, can also show up when processing interrupts: In order to have a properly defined state when the processing of an interrupt begins, all instructions for which the program counter has been incremented must be allowed to complete execution before the interrupt-processing is initiated if no arrangements have been made for a rollback; the number of such instructions (and, therefore, the delay involved) is proportional to the number of stages between the stage at which the control point is located (if the program counter is altered there) and the end of the pipeline, as is the likelihood that the instructions will change machine state to a non-trivial degree. Many current high-performance machines have special provisions for precise interrupts — although there are a few, e.g. the Silicon Graphics TFP, for which imprecise interrupts is norm — so this in itself is not an issue here. But, as we shall see in Chapter 7, the mechanisms that are used to facilitate precise interrupts also tend to be used to recover from branch misprediction (and associated speculative processing), and the two types of "exceptions" can therefore be closely related, regardless of whether the program counter is altered at the control point.

The best placement for the control point is not easily determined, as there is a push-and-pull effect involved: On the one hand, the delay encountered (after fetching the target instructions) in filling the pipeline after an unconditional branch, or after a conditional branch for which the condition is known, is determined solely by the number of stages between the first stage of the pipeline and the control point; minimizing this delay suggests that the control point should be at an early stage or that the control point should be partitioned into two parts, with an early part used for unconditionals. On the other hand, for a conditional branch for which the condition is not known, if the instructions following the branch are from the correct path, then the delay encountered is determined by the number of stages between the control point and the stage at which the branch condition is evaluated; minimizing this delay suggests that the control point should be at a late stage. (It should be noted that a conditional branch instruction will be followed into the pipeline by other instructions only if there is some sort of target-address prediction employed; even simple "proceed-and-squash", as described above is, in effect, a prediction — that the branch will not be taken. If there is no branch-target prediction, then the pipeline latency for a conditional branch is proportional to the length of the pipeline up to the condition-evaluation stage; usually, but not necessarily, this is the length of the entire pipeline.) But instructions that are conditionally (i.e. speculatively) processed might non-trivially change machine-state, and the number of such instructions and the degree to which they can change machine-state, as well as the cost of saving and restoring the state if the prediction is incorrect, increases as the control point gets farther from the beginning of the pipeline; so this suggests an early control point. Also, for any type of branch instruction, having a late control point means that the addresses of the target can be calculated well before the branch has to be effected. A late control point is in fact required if some preliminary processing is required to compute the target address: for example, when operand-accessing is necessary in indirect addressing or addressing relative to an arbitrary base. Table 4.1 is a rough summary of the preceding discussion.

Table 4.1.   Pipeline latency relative to control point

Branch Type	Condition Status	Next Instructions	Delay (cycles)
conditional	known condition	correct sequence	0
	known condition	incorrect sequence	$k+1$
	unknown condition	correct sequence	m
	unknown condition	incorrect sequence	$k+m+1$
stalling	—	—	$k+m+1$
unconditional	—	—	$k+1$

m = number of stages from control point to condition-evaluation stage
k = number of stages before control point

The control point therefore appears in different positions, according to the design of each particular machine: For example, in the Cray and CDC super-computers, in which performance is mainly derived from heavy pipelining in the multiple execution units, but the instruction preprocessing stages are lightly pipelined, the control point is at the beginning of the pipeline; similarly, in machines without sophisticated branch-processing hardware, the control point typically is in the FETCH/DECODE stage. In many RISC machines with short pipelines, the control point is at the last stage. Other examples: The MIPS R4000 has its control point halfway down the pipeline (at the fourth of eight stages), which means that only three instructions can be in the pipeline behind a branch. None of these instructions can change machine-state in a none-trivial way — they are at the FETCH, ISSUE, and READ-REGISTER stages — and in the worst case only two of the three instructions have to be discarded.[3] In the MIPS R10000, on the other hand, the control point is at the second stage, with the preceding stage corresponding to a delay slot. In the Manchester MU5 (Section 1.5.3), the control point is about midway in the complete processor pipeline — at the end of the primary pipeline and the beginning of the secondary pipeline — although in this case several instructions behind a branch can cause non-trivial changes in machine-state. And in the DEC Alpha 21264, the control point is in the sixth stage.

It should, however, be noted that the figures in Table 4.1 are best-case figures that deal with just pipeline stages and which, therefore, do not take other delays (e.g. memory access or buffering) into account. So, for example, in the DEC Alpha 21264, $k = 6$, but the actual branch latency ranges from 7 cycles to 12 cycles. We should also remark that a combination of hardware and software techniques can be used to give smaller *effective* delays than those in Table 4.1. For example, in an architecture that uses explicit condition codes for conditional branching, if a sufficient number of unrelated instructions can be found to go between the condition-setting instruction and the condition-testing instruction — a technique known as *branch spreading* — then the condition will be known by the time the branch instruction gets to the control point, and m will effectively be zero[4]. A number of related techniques are discussed below; indeed, almost all of the chapter is devoted to ways of effectively reducing the delays in Table 4.1.

In a number of current superscalar machines, such as the Intel Pentium, the AMD-K6, etc., branches are executed by a unit that works in parallel with the other execution units; therefore the control point is usually at a late stage, unless branch instructions can be detected and processed quite early. These machines therefore require some means of fetching instructions and filling the pipeline before the outcome of the branch is known; moreover, the basic problem is aggravated by the need to be able to pre-process and execute several instructions in parallel. Consequently, most of these machines implement very sophisticated target-address prediction techniques.

4.2   LATENCY REDUCTION BY INSTRUCTION BUFFERING

In this section we shall examine the use of simple hardware instruction-buffers to reduce the fetch latency. The sole aim of these is balance the flow of instruc-

---

[3]The third is in a "delay slot" — see Section 4.3.
[4]A conditional branch is said to be *resolved* if the outcome is known beforehand.

tions between a memory unit with a high access time and a pipeline with a relatively small cycle time. (Although the memory access time might be high, techniques such as multiple-bank interleaving can be used to provide the high bandwidth required to fill such a buffer.) In some cases, these systems also incorporate simple means (i.e. none that involves extra hardware costs) that attempt to reduce the deleterious effects of branch instructions by taking advantage of the regularity in loop-closing branches. In what follows, we shall categorise the different types of buffer, give a few examples, and then compare the different approaches.

4.2.1   Types of instruction buffer

The simplest type of instruction buffer is a straightforward first-in/first-out (FIFO) queue. The sole purpose of such a buffer is to support prefetching, and the buffer is therefore known as a *prefetch buffer*. One pointer, Head, indicates the head of the queue, and another, Tail, points to the tail. Prefetching is used to insert instructions at one end of the queue, with Tail incremented accordingly, while the pipeline takes instructions from the other end, incrementing Head accordingly. The arrangement has some usefulness in branching: if the branch target is already in the buffer, then an access to memory is not needed. This condition can be detected in one of the following two ways. Suppose the queue consists of N lines. The first method is that if Head and Tail point only to the queue entries, and the other bits are the same as the corresponding bits of the program counter — i.e. they can take values only in the range $0, 1, \ldots, N - 1$ — then a check is made determine whether the branch target is within $N - 1$ words of the program counter and within the range delimited by Head and Tail. And the second method is that if Head and Tail hold instruction addresses — that is, the lowermost $\log_2 N$ bits of each point to queue entries — then a comparison is made of the target address with the contents of both Head and Tail, with the lowermost $\log_2 N$ bits excluded, and the displacement from Head locates the target instruction.

The basic prefetch buffer partially solves the fetch-latency problem, but it is not very effective, in that it does not take repeated use of instructions into account and therefore requires more memory bandwidth support than would otherwise be the case. Another of the buffer's drawbacks is that that because instructions are entered in strict sequence, it must be flushed and refilled whenever it does not contain the target instruction; instructions may therefore be wastefully prefetched. For these reasons, very few machines, use a prefetch buffer on its own, but it may be, and frequently is, used as a supplement to other structures.

The type of branching structure that is easiest to take advantage of is branching in a loop. So a straightforward way to address (partially, at least) the drawbacks above of the prefetch buffer is to arrange matters so that in the event of a loop, repeated accesses are made to the same instructions in the buffer, and no accesses to memory are then necessary. The resulting buffer is then a *loop-catching buffer*. Such a buffer is an improvement on the simple prefetch buffer, but its effectiveness is nevertheless limited by the built-in assumption that a trapped loop consists of a contiguous segment of code that lies between the address of the branch instruction and its target address. This means that unless the buffer is quite large, it may be of little use with loops consisting of non-contiguous segments of code; in particular, it is possible that the actual code of the loop is small enough to fit in the buffer even though the difference between the two addresses might be larger than the buffer size. Extending the system to deal with non-contiguous loops requires that lines of the buffer be individually addressable: an obvious extension of the second addressing implementation outlined above will do this and, in the extreme, leads to an instruction cache. The instruction cache is useful in the solution of the problem of fetch latency, and it also reduces the overall memory-bandwidth requirements, but it is of little help with the pipeline latency induced by a branch instruction. For the pipeline-latency not to be problematic, some means, such as branch-target prediction is required in order to make sure that the buffer or cache always contains such instructions as are required to keep the pipeline free of "empty bubbles".

4.2.2 Case Studies

In the remainder of this section we shall examine four examples of simple instruction-buffering systems. The first three, which are all historically related, are the systems used in the CDC 6600 (the ancestor of high-performance machines from Control Data Corporation and Cray Research), the Cray-1 (from which all Cray machines are derived), and the CDC 7600 (which has the type of buffer used in later machines from CDC and ETA) [14, 24, 111]. The Cray system has some similarities with the CDC 7600 system, but it does not allow address resolution to the same extent and is therefore less sophisticated. The fourth example is the set-associative instruction cache used in the IBM RS/6000; this machine differs from the other examples in that it is superscalar, and the instruction-buffering system therefore has to provide several instructions in each pipeline cycle [47]. The last example is the instruction buffering system of the DEC Alpha 21164 microprocessor, which is an example of the state-of-the art in microprocessor implementation [34].

4.2.2.1 CDC 6600, Cray-1/S, and CDC 7600 instruction buffers. The CDC 6600 Instruction Stack has the organization shown in Figure 4.1. It consists of eight lines, each of which holds at least two 30-bit instructions and at most four 15-bit instructions. During normal instruction execution, the lines are filled in one at a time, with each new entry (at the bottom of the stack) causing older entries to move one slot up; this takes place at a rate that matches the pipeline cycle time. The contents of the stack are declared void whenever a branch takes place to an instruction that is not in the stack.

To keep track of the current valid contents of the stack, two pointers are maintained: the Depth pointer indicates the last valid entry made to the stack, and the Locator pointer indicates the line from which instructions are currently being taken. Thus when an out-of-stack branch occurs Depth will have the value zero, and when the stack is full Depth will have the value seven.

Whenever a loop is small enough to be wholly contained in the buffer, the fetching of instructions from the main memory ceases. This condition is detected by carrying out one or two tests: the first is to check that the target instruction is within seven words of the current value of the program counter; if it is, then the other test (which involves comparing the contents of Depth and Locator) is to determine if the target is validly within the stack. This arrangement works well for small loops but fails for large loops and for loops that cover disjoint and widely separated pieces of code; in the latter case, it does not work if the addresses that delimit the loop differ by more than the buffer size, even if the actual loop size is smaller.

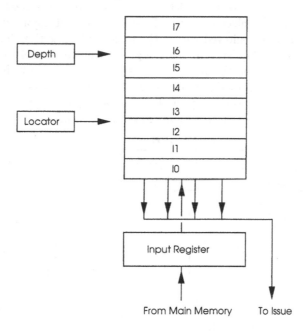

Figure 4.1.   CDC 6600 instruction stack

The supercomputers produced by Cray Research all use a similar instruction buffering system, first implemented in the Cray-1 computer. The Cray-1/S system, which has the organization shown in Figure 4.2, consists of four instruction buffers and three associated registers. Each line in each of the buffers holds a single word of four 16-bit *instruction parcels*. (An instruction consists of one or two parcels.) The Current Instruction Parcel (CIP) register holds the first parcel of the next instruction to be issued and is filled, via the Next Instruction Parcel (NIP) register, with successive instruction parcels from the buffers. In the case of a two-parcel instruction, the Lower Instruction Parcel (LIP) register holds the the lower half of the parcel pair. The Program Address (PA) register holds the address of the next instruction that is to enter the NIP register: the lower two bits of the PA register specify the parcel within a word, and the remaining bits specify a main memory address; unless a branch occurs, the value in PA incremented at every cycle.

Each of the four instruction buffers holds sixty-four instruction parcels — four in each of sixteen slots. Associated with each buffer is a 18-bit Beginning Address Register (BAR) that holds the the first eighteen bits of the memory addresses of the instructions in that buffer. During every cycle, the high-order bits of the PA register are simultaneously compared with the contents of all of the BARs. (The system may therefore be viewed as a fully-associative cache with a very small number of large lines.) If a match occurs, then the instruction is in the buffers, and the matching BAR identifies which buffer. If there is no match, then new instructions, including the required one, are loaded into one of the buffers, at a rate of four 64-bit words per cycle; the next buffer to be filled in such a case is selected using a First-In-First-Out algorithm. Branches, in either direction, are allowed within the buffer, which is therefore capable of holding small loops. The system is an improvement on that of the CDC 6600, in that the entire buffer is not accessed as a single unit, but the large-grain partitioning of of the buffer (i.e. the small number of BARs) means that the extent to which several loops, especially those of disjoint code segments can be handled is not as great as in the type of system used in, say, the CDC 7600.

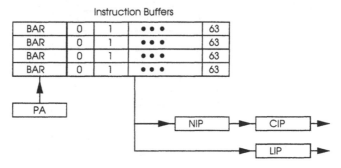

Figure 4.2.   Cray-1 instruction buffers

The CDC 7600 system is basically an extension of the CDC 6600 system. The instruction buffer, called the Instruction Word Stack (IWS), consists of twelve lines, each of which holds a single word; a word is sixty bits wide and contains a mixture of 30-bit and 15-bit instructions, each in one or two 15-bit *parcels*. The IWS is filled by lookahead (two words ahead of the current value of the program counter), and during normal sequencing, each new entry causes the older entries to move one line up in the stack; the stack is therefore similar to the CDC 6600 system in these respects. The main difference between the CDC 6600 and CDC 7600 systems lies in the Instruction Address Stack (IAS) of the latter. The IAS consists of twelve registers, each of which holds the main-memory address of a word in the IWS. As the program counter is advanced, each new value is concurrently (associatively) compared with all of the entries in the IAS. (The IAS-IWS combination is therefore, essentially, also a fully-associative instruction cache with a small number of lines.) If there is a match, then the corresponding word from the IWS is read into the Current Instruction Word (CIW) register, which holds the next instruction parcels to be processed. If there is no match — that is, if an out-of-stack branch has occurred — then new words are requested from memory, and processing continues after they arrive. The system can therefore completely buffer small loops, removing the need to access memory in such cases. Moreover, unlike the CDC 6600 system, the fact that the IWS lines are addressed individually (by the contents of the IAS) means that there is no difficulty in holding a loop of disjoint code segments, as long as the total loop-size is sufficiently small; also, the entire buffer does not have to be flushed when an out-of-stack branch occurs. The design in fact allows more than one loop to be held in the buffer. The main difference between the CDC 7600 buffer and that of the Cray-1 is in the granularity of partitioning: each unit addressed by a BAR in the Cray-1 holds 64 parcels, in contrast with 4 parcels per IAS register in the CDC 7600.

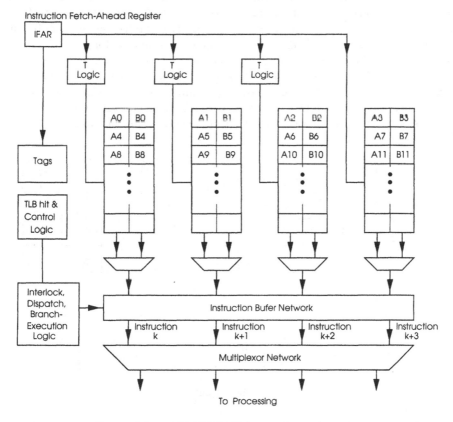

Figure 4.3.   IBM RS/6000 instruction cache unit

4.2.2.2 IBM RS/6000 instruction cache unit.

Instruction buffering in the IBM RS/6000 processor is carried out in the Instruction Cache Unit, which consists of a cache to hold instructions, a branch-instruction processor, instruction-dispatch logic, and some interrupt-handling logic. The processor is superscalar, and the instruction-supply unit therefore has to be capable of supplying several (four) instructions in each cycle.

The organization of the Instruction Cache Unit is shown in Figure 4.3. The instruction cache is an 8Kbyte, two-way set-associative cache, with a line width of 64 bytes (16 instructions). It is partitioned into four arrays, organized in such a way that that each array can provide the rest of the pipeline with one instruction in each cycle: Each array is partitioned into two parts (A and B in Figure 4.3) that correspond to the two lines in a set, and instructions from a sequential stream are interleaved across the arrays, with instruction i going into array i **mod** 4. The logical layout is thus as shown in Figure 4.4. In order to be able to fetch a sequence of four instructions that lie in the same line, the row address to the arrays may have to be incremented, depending on how the four instructions are aligned relative to the address of the first one: to fetch instructions 0, 1, 2, and 3, all arrays get the same row address; to fetch instructions 1, 2, 3, and 4, the row address for array 0 must be incremented, as instruction 4 lies in a different row from the others; and so forth. The T-logic units provide the required row-increment and row-select functions. The tag directory is split into two parts — one corresponding to the even-numbered sets and another corresponding to the odd-numbered sets — that are searched in parallel, thus providing concurrent access to two consecutively numbered sets, as is required when four consecutive instructions are in two sets.

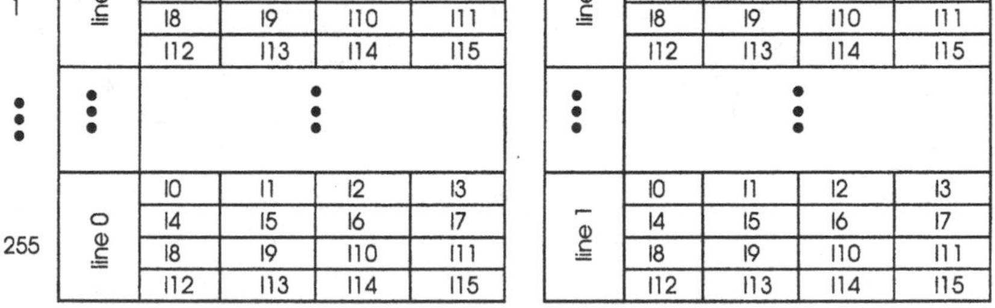

		A						B			
Set	Tag	Array 0	Array 1	Array 2	Array 3		Tag	Array 0	Array 1	Array 2	Array 3
0	line 0	I0	I1	I2	I3		line 1	I0	I1	I2	I3
		I4	I5	I6	I7			I4	I5	I6	I7
		I8	I9	I10	I11			I8	I9	I10	I11
		I12	I13	I14	I15			I12	I13	I14	I15
1	line 0	I0	I1	I2	I3		line 1	I0	I1	I2	I3
		I4	I5	I6	I7			I4	I5	I6	I7
		I8	I9	I10	I11			I8	I9	I10	I11
		I12	I13	I14	I15			I12	I13	I14	I15
⋮	⋮		⋮				⋮		⋮		
255	line 0	I0	I1	I2	I3		line 1	I0	I1	I2	I3
		I4	I5	I6	I7			I4	I5	I6	I7
		I8	I9	I10	I11			I8	I9	I10	I11
		I12	I13	I14	I15			I12	I13	I14	I15

Figure 4.4.   Instruction layout in RS/6000 cache

The cache can return four instructions per cycle in the best case and just one in the worst case. Figure 4.4 shows that for a given sequence of four instructions, in thirteen out of every sixteen cases the instructions will be in the same line, and in the remaining three cases they will span two lines. In the latter three cases, the possibilities are that the first three instructions are in the line first accessed, or that two instructions are in that line, or that it is just one instruction that is in that line. So the average number of instructions returned in each cycle is $(13/16) \times 4 + (1/16) \times 3 + (1/16) \times 2 + (1/16) \times 1 = 3 \cdot 625$.

Branches in the machine are processed by a special unit that overlaps the execution of branch instructions with those of other instructions and integrates branch-processing with instruction-buffering. Unconditional branches are processed in a straightforward manner and do not affect performance. In the case of conditional branches, it is assumed that the branch will be taken: the instructions from the target address are conditionally issued and are then squashed or completely executed when the branch condition has been evaluated. Overlapping the execution of branch instructions with that of other instructions masks most of the latencies involved but requires that branch instructions be detected sufficiently early. Consequently, the branch-processing logic scans the arrays to detect branch instructions well in advance of when they actually need to be executed and, where possible, determines branch conditions and target addresses in advance. The raw bandwidth of the cache is therefore greater than what is nominally required by the execution pipelines.

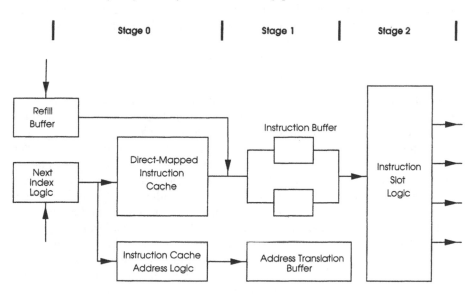

Figure 4.5.   Instruction preprocessing in DEC Alpha 21164

4.2.2.3   DEC Alpha 21164 instruction-fetch and buffering.

The DEC Alpha 21164 is a second-generation implementation of the DEC Alpha architecture (Section 1.5.4). The processor consists of a superscalar pipeline in which up to four instructions can be issued in each cycle. The instruction-preprocessing stages of the pipeline are shown in Figure 4.5. In stage 0, the Instruction Cache (an 8Kbyte, 32byte-block, direct-mapped cache) is accessed; and in stage 1, a new address is selected (between the next sequential address and a branch-target address) for the next access to the cache, the instructions fetched in stage 0 are entered in the Instruction Buffer, and branch instructions are decoded. During normal operation, the cache returns a four-instruction block (16 bytes) in each cycle. Each instruction is accompanied by five bits of pre-decoded information that are used in stage 1 for processing branches and in stage 2 for instruction issue. An instruction-prefetcher that operates concurrently with the cache tries to keep it full at all times; the prefetcher generates virtual addresses (of 32-byte blocks) that are translated in the Translation Buffer and then sent to the memory, which returns instructions into the 4-block Refill Buffer. Instructions then move, when needed, from the Refill Buffer into the Instruction Buffer. The cache is updated during the transfer between the buffers, and if the data movement creates an empty slot in the Refill Buffer, then a request to memory is made to fill the slot. Instructions are initially buffered only in the Refill Buffer, instead of being entered in the cache as well, in order to avoid unnecessary replacement of valid blocks: a miss in the cache results in a check in the Refill Buffer for the required instruction; for this check, each slot of the buffer holds an address and has a comparator. Instruction-fetching from memory begins only when the required instruction is not in the cache or the Refill Buffer and stops when a branch instruction is decoded or as soon as the required instruction has been placed in the cache.

4.2.3 Comparison of instruction buffers

The different types of buffer may be classified and analysed according to the granularity in which the storage is partitioned: The largest grain of partitioning yields the simplest type of buffer, in which the entire storage is accessed as a single unit, and is exemplified by the queue structures used in the CDC 6600 and several early low-performance microprocessors from Intel and Motorola. At the second level of partitioning, the buffer is divided into a very small number of moderately-sized sections that can be accessed independently of each other; examples of this type of buffer are those in the Cray machines. And at the third level, the buffer is partitioned into many small sections that can be accessed independently; this yields an instruction cache that may be direct-mapped, set-associative, or fully-associative and is the most common in today's high-performance machines, with the direct-mapped cache increasingly used (for the performance reasons given in Chapter 3).

The choice of which type of buffer to implement is a matter of trading off cost and performance. Assuming the same amount of storage, the large-grain buffer, which requires the least number of addressing registers, is the cheapest; the medium-grain buffer, which requires a few more registers for addressing, is next in cost; and the fine-grain buffer, which requires the most sophisticated addressing is the costliest.

In addition to a straightforward reduction of the fetch latency, all three types of buffer can capture loops and can therefore be useful with some branches. The large-grain buffer does, however, have the drawbacks that it is limited to one loop at a time, that the difference between the addresses delimiting the loop must be no larger than the buffer size (even though the loop itself may be smaller than that), and that it must be flushed when an out-of-buffer branch takes place. The medium-grain buffer is an improvement, since it can capture as many loops as there are sections, and it is flushed on a section-by-section basis; nevertheless, the grain size limits its effectiveness when there are many small loops or highly unstructured code. Consequently, while the medium-grain buffer can be quite effective in a machine intended to run mainly scientific code (in which a typical phase of computation involves a long time spent on the same code, over structured data), we should expect that it will not be as effective with general-purpose computation: note the data in Tables 4.3 and 4.6, which indicate that there is relatively not much branching in scientific code and that most branches in such code are taken; moreover, in vector-pipeline machines, such as the Crays, a large number of elementary operations will be encoded into a small number of vector operations, thus reducing the overall required instruction-delivery rate. The small-grain buffer is the best in these respects, and we may therefore say that performance and cost increase with decreasing granularity. Additional discussion of basic instruction buffering will be found in [67].

4.2.4 Summary

We have discussed the use of hardware buffers (placed between memory and processor) to sustain the flow of instructions into the processor pipeline. The basic rationale for such buffers is that even though the memory may have a high access time (relative to the processor cycle time), with sequential addresses, the buffers can (through the use of memory interleaving) be filled at a high rate and can therefore supply the pipeline at the required rate. The different types of buffer vary mainly in the number of independent sections into which the storage is partitioned, and both performance and cost increase with the number of such sections. In current high-performance machines, the need to match instruction access time to processor cycle times and to be able to deliver several instructions in each cycle means that the buffer used is likely to be a direct-mapped cache, or perhaps set-associative, with partial or full virtual addressing and based on interleaving techniques.

The main limitation of all the buffers is that they mainly reduce only the fetch latency, although being able to prefetch instructions implies that at least part of the address latency must also be masked. Therefore, for a complete solution to the branching problem, such buffers must be supplemented with the use of additional techniques that reduce or mask the other latencies. This is the approach taken in high-performance machines, and in what follows we shall discuss some of these other techniques.

4.3 STATIC INSTRUCTION-SCHEDULING

A number of simple pipelined machines rely solely on software techniques to deal with the problem of branching; and in very-high-performance machines, such techniques are also used to supplement more aggressive hardware-implemented techniques. In the use of software techniques, the compiler or other system software arranges for code to be generated and placed in a manner that is more conducive to pipelining than straightforward code or the normal flow of events would dictate. Such techniques may be supported by or predetermined by the underlying architecture, or they may be purely compiler techniques. The term *static scheduling* indicates the determination of execution order before run-time; in contrast, *dynamic scheduling* techniques are used for run-time reordering (Chapter 5).

One static scheduling technique that came into widespread use with the advent of RISCs is *delayed branching*, in which code is re-arranged so that a branch is effected some time after what its position indicates in normal code. This is organized in such a way that all instructions entering the pipeline are guaranteed to complete and the pipeline executes one (ideally, but not necessarily, useful) instruction in each cycle. The goal is to try and execute a number of useful instructions while the direction of the branch is being resolved and the target instructions fetched.

Table 4.2.   Delayed branching

Instruction Address	Normal Branch	Delayed Branch	Optimized Delayed Branch
100	LOAD X, A	LOAD X, A	LOAD X, A
101	ADD 1, A	ADD 1, A	JUMP 105
102	JUMP 105	JUMP 106	ADD 1, A
103	ADD A, B	NOP	ADD A, B
104	SUB C, B	ADD A, B	SUB C, B
105	STORE A, Z	SUB C, B	STORE A, Z
106		STORE A, Z	

The basics of delayed branching are as follows. Suppose the delay incurred on a branch is equivalent to k pipeline slots; that is, k new instructions could be admitted into the pipeline in the duration incurred from processing a branch. Then if k instructions could be found to be processed (unconditionally) while the branch target was being fetched, the pipeline would be kept occupied at all times, and the branching would then not affect performance. In essence, the effect of the branch is delayed by k slots, known as the *delay slot*. Table 4.2 illustrates this technique for a delay slot of one cycle: the sequence of code in the first column is the normal code; the code in the second sequence is what is obtained by delaying the effect of the branch by one cycle, through the insertion of a *no-operation* (NOP) instruction; and the code in the last column is the result of removing the NOP (i.e. *filling* the delay slot with a useful instruction). The instruction used to fill the delay slot could also have been taken from code preceding the branch: being able to take the filler instructions from before or after the branch instruction increases the likelihood of finding a suitable instruction. Additionally, at least one study has shown that the effectiveness of delayed branching can be increased by using it in conjunction with a simple target-address prediction strategy [108]. In *delayed-branch-with-squashing*, thetarget instructions in the delay slot are executed only if the branch is taken as predicted by the compiler; otherwise, they are squashed [19].

Evidently, much of the effectiveness of the delayed branching technique depends on the extent to which the NOPs can be removed and replaced with useful instructions, which requires some sophistication in the compiler and also depends on the machine architecture. The removal of NOPs is not particularly difficult for unconditional branches but is problematic for the conditionals: for example, in the Berkley RISC I, the designers quote a 90% success figure for unconditionals but only 25% for conditionals, and for the IBM 801 the figures are worse, with only 60% success for the unconditionals [94, 100]. It is also hard to fill the delay slot if it is larger than one: in one study, it was found that for a two-slot delay, the first slot could be filled 70% of the time, but the second slot could be filled only 25% of the time [76]. Many implementations of delayed branching, especially those in the early RISCs, have used a delay slot of only one cycle; the Cheetah, the Power implementation that led to the IBM RS/6000, is one of the few examples of the use of a larger delay slot — four [47, 88].

A straightforward way to extend the technique of delayed branching is to remove the requirement that a fixed number of instructions be executed before the branch is effected; this then eliminates the need for NOPs. One such extension is *branch preparation*, in which a PREPARE-TO-BRANCH (PTB) instruction is used to alert the pipeline of branching further down in the instruction stream [41, 129]. The branch itself is effected in the normal manner, but the PTB instruction indicates how far away the *effective* branch is (and what its target is) and, therefore, the number of instructions that must be processed, irrespective of whether the branch is taken or not. Essentially, the PTB realizes delayed branching with a variable-length delay slot.

We have seen in Section 4.1 that part of the pipeline latency for a conditional branch depends on the number of stages between the control point and the condition-evaluation stages or. This, however, is only so for those cases where the branch instruction has to wait for a condition to be evaluated in the execution units or some other late stage. Now, in most normal code in which conditional branching is accomplished with two instructions, the branch instruction typically immediately follows the condition-setting instruction, so the delay is inevitable, unless some special action is taken to deal with it. If, however, some code could be found to go between the branch and the condition-setting instruction, then by the time the branch got to the control point the condition would already have been evaluated and the delay therefore eliminated or at least reduced. Deliberately trying achieve a maximum separation between the two instructions is another partial solution to the branching problem and is known as *branch spreading*.

Another partial software/hardware solution to the branching problem is the *branch folding* that is implemented in the CRISP and Toshiba TX3 microprocessors [29, 89]. CRISP has an instruction cache that holds decoded instructions and in which every instruction has appended the address of the next instruction to be processed. This means that every instruction can perform a transfer of control and obviates the need for the execution of explicit branch instructions: during decoding, when a non-branching instruction is followed by a branching instruction, the two are combined ("folded") into a single instruction, thus eliminating the time that would otherwise be required to execute an explicit branch. To fold conditional branches, an extra field is used to hold the alternate value of the program counter. The execution then assumes that the alternate program-counter value will not be needed, but it is carried along the pipeline in case this assumption is not correct; when the condition is known, either execution continues, or the pipeline is flushed and the alternate program-counter value used. In superscalar machines, an effect similar to that of branch folding can be achieved by arranging code so that the execution of a branch instruction (in a dedicated *branch processor*) is overlapped with the execution of computational instructions in the arithmetic units. Thus, for example, in the IBM RS/6000, it is possible to have three instructions executed in parallel: one in the branch unit, one in the fixed-point arithmetic unit, and one in the floating-point arithmetic unit [88]; and in the AMD K6, up to six other instructions can be processed concurrently with a branch instruction [4, 52]. If we take the view that a branch instruction changes the flow of control but does no "useful" work, then an additional measure for the cost of a branch is the number of cycles taken in its execution. Where the execution of a branch instruction is completely overlapped with that of useful instructions, its execution cost will be 0, whereas it would otherwise be at least 1; as an example, the IBM RS/6000 has an average branch cost of 0.5 cycles.

All of the techniques discussed above can be, and most have been, used as supplements to other techniques that are described below. Although they have recently become very popular, none is entirely new. For example, delayed branching and branch folding can be traced to old microprogramming techniques, and branch spreading has long been used in a number of machines as a supplement to hardware solutions to the branching problem. The limitations of these techniques include the following: One, they require very sophisticated compilers, and in any case there are instances where code cannot be manipulated in the manner indicated. Two, it may be hard for a programmer to understand code that has been generated by a compiler that manipulates code in such ways, which means that debugging can be difficult. Three, they complicate the handling of interrupts, especially if the delay slot is large. Four, none is very effective with superpipelined machines — the longer the pipeline, the larger the delay slot and the harder it is to find enough useful instructions to fill it — or, for similar reasons, with superscalar machine. The last of these is especially important, as in the future more machines will be superpipelined or superscalar to a high degree; indeed, this trend is already visible in commercial microprocessors. One important implication of this is that it can be harder to maintain complete software compatibility between different implementations of the same architecture as changes are made for performance. A technique such as delayed branching is no longer as important as it once was, but there are still a number of recent machines (e.g. Hitachi's SuperH RISCs) that use it as the main solution to the branching problem and other recent ones (e.g. the MIPS R10000 and the HP Precision Architecture) that retain it solely to facilitate compatibility with preceding architectures and implementations [54, 121].

4.4 BRANCH PREDICTION

In this section we shall discuss the use of target-prediction techniques to reduce the three branching latencies. The basis of many of these techniques is that they reduce the address latency in such a way that the fetch and pipeline latencies are also (directly or indirectly) reduced or masked. The address-latency problem — that of determining the target address with minimal delay and, ideally, early enough to overcome the fetch latency, or the pipeline latency, or both — is mainly a problem only for conditional branches. For unconditional branches, the basic problem is easily solved if the control point is early and branch instructions can be detected early enough; hence in implementations such as the IBM RS/6000, the instruction cache is scanned to detect branch instructions before they enter the pipeline proper. For conditional branches, on the other hand, there seems to be an unavoidable delay — either between the start of the pipeline and the control point, or between the control point and the condition-evaluation stage (and, therefore, the determination of the target of the branch), or both. The best solution in this case is to try and predict the the outcome of the condition-evaluation early enough. Such prediction would then allow processing to continue and thus reduce or eliminate the inherent delay (if the prediction is correct). Several current high-performance machines employ such prediction techniques, and in very aggressive implementations this is be extended to allow conditional execution of instructions beyond several unresolved branches.

In what follows we shall look at such techniques (mostly those implemented in hardware) that deal with the conditional branches by predicting the result of the condition-evaluation. The branch prediction methods can be classified into three broad groups: *static*, in which execution-time behaviour is not taken into account, *dynamic*, in which branching behaviour is "analysed" and used *during* program execution, and *semi-static*, which lies somewhere between the static and dynamic methods.

Table 4.3.   Proportion of taken branches

Branch	IBM System/370				DEC PDP-11	CDC 6400	Avg.
	CPL	BUS	SCI	SUP			
taken	0.640	0.657	0.704	0.540	0.738	0.778	0.676
not-taken	0.360	0.343	0.296	0.460	0.262	0.222	0.324
P_b	0.317	0.189	0.105	0.376	0.388	0.079	0.242

CPL=compiler, BUS=business, SCI=scientific, SUP=supervisor
P_b=probability that any instruction is a branch

SPEC program	P_u	P_c	P_{bt}	P_{ct}
spice	0.093	0.125	0.538	0.196
doduc	0.020	0.094	0.630	0.551
nasa7	0.000	0.166	0.994	0.993
matrix300	0.001	0.198	0.993	0.993
fpppp	0.005	0.016	0.575	0.450
tomcatv	0.000	0.059	0.993	0.993
gcc	0.041	0.189	0.635	0.556
espresso	0.071	0.193	0.538	0.369
li	0.062	0.165	0.601	0.450
eqnott	0.021	0.305	0.455	0.406

P_u=probability of unconditional branch
P_c=probability of conditional branch
P_{bt}=probability branch is taken
P_{ct}=probability conditional branch is taken

4.4.1   Static and semi-static prediction

An example of simple static branch prediction is to predict that all branches will be taken. This prediction can be quite effective, given that estimates of the proportion of taken branches in sample programs (see Table 4.3) have been as high as 68–70% [69, 90, 91]. The obverse is to predict that all branches will not be taken, which is just the strategy, described at the beginning of the chapter, of continuing on the straight-ahead path. We might not expect this to be as effective as "predict taken" if most branches are indeed taken, but the effectiveness of many static methods can be improved by the use of compile-time techniques that re-arrange code to increase the likelihood that branches will be taken in the direction predicted [8, 10]. One advantage of predicting "not-taken" is that the target address need not be computed, unless the prediction is incorrect.

Another example of static prediction is to base the prediction on the opcode: Various studies have found that, for particular machine architectures and work-

Table 4.4.   Frequencies of branches by opcode

	IBM System/370				DEC PDP-11		CDC 6400	
	CPL	BUS	SCI	SUP	CPL		SCI	
BR, B	0.222	0.243	0.254	0.138	JSR	0.111	RJ	0.049
BAL	0.056	0.036	0.013	0.036	SOB	0.008	JP	0.017
BALR	0.036	0.050	0.079	0.065	BGET	0.113	XJ	0.560
BCT	0.024	0.013	0.027	0.016	BVCS	0.030	EQ	0.157
BCTR	0.022	0.050	0.006	0.019	BHSL	0.031	NE	0.199
BXH	0.004	0.000	0.000	0.000	BNEQ	0.278	GE	0.000
BXLE	0.032	0.000	0.188	0.003	RTS	0.074	LT	0.003
BC	0.544	0.521	0.318	0.674	JMP	0.190	SYS	0.015
BCR	0.051	0.081	0.112	0.034	BR	0.162		
EX	0.009	0.005	0.003	0.005	TRAP	0.002		
SVC	0.000	0.001	0.000	0.001				
LPSW	0.000	0.000	0.000	0.005				
MC	0.000	0.000	0.000	0.005				

loads, branches of a given type tend to behave in a mostly predictable manner. For example, one study of CDC Cyber 170 programs showed that BRANCH-IF-NEGATIVE, BRANCH-IF-EQUAL, and BRANCH-IF-GREATER-THAN-OR-EQUAL were usually taken, and the results (see Tables 4.4 and 4.5) of another study, for the IBM 370 are similar [09]. The VAX 9000 is an example of a machine that uses opcode-based prediction, in addition to dynamic prediction [85].

A third example of static prediction is to make the prediction according to the direction of the branch. Thus, for example, since most loop-closing branches tend to branch backwards, one such strategy would be to predict that all forward branches will not be taken and all backward branches will be taken. This type of prediction is used in the National Semiconductor NS32532 microprocessor, in which its combination with opcode-based prediction has a reported accuracy of about 71%, in the DEC Alpha 21164, as a supplement to dynamic prediction, and in the Hewlett-Packard PA7100 [5, 34, 58]. The PowerPC 601 also predicts that backward branches are taken and forward ones are not, but the compiler is given the option of reversing this by setting an appropriate bit in the branch instruction's opcode field [7].

A fourth possibility for static branch prediction is to predict according to the distance of the target instruction from the branch instruction: if the distance is smaller than some pre-set value, then predict "taken"; otherwise, predict "not-taken". The rationale for this type of prediction is that loop-closing branches tend to have branch-distances that are smaller than those for other types of branches and are usually taken.

Static prediction can also be used to increase the effectiveness of delayed branching, as well as of other prediction methods discussed below; in general, in situations in which a branch is mostly taken, static prediction works well and

Table 4.5.   Probabilities of taken branches by branch type

| | IBM System/370 | | | | DEC PDP-11 | | CDC 6400 | |
	CPL	BUS	SCI	SUP				
BR, B	1.000	1.000	1.000	1.000	JSR	1.000	RJ	1.000
BAL	1.000	1.000	1.000	1.000	SOB	0.448	JP	1.000
BALR	0.659	0.555	0.850	0.531	BGET	0.330	XJ	0.604
BCT	0.584	0.899	0.857	0.713	BVCS	0.155	EQ	1.000
BCTR	0.007	0.173	0.000	0.027	BHSL	0.496	NE	1.000
BXH	0.404				BNEQ	0.495	GE	0.848
BXLE	0.865	0.994	0.865	0.522	RTS	1.000	LT	0.000
BC	0.462	0.571	0.342	0.415	JMP	1.000	SYS	1.000
BCR	0.539	0.348	0.647	0.584	BR	1.000		
EX	1.000	1.000	1.000	1.000	TRAP	1.000		
SVC	1.000	1.000	1.000	1.000				
LPSW			1.000					
MC			1.000					

is inexpensive. Although static prediction methods can work well in certain cases, on the whole they are less accurate than dynamic less, as in most cases the direction of a branch tends to be influenced by past behaviour or by the behaviour of other (related) branches. The effectiveness of a static method can therefore be very dependent on the particular program executed. Semi-static methods partially deal with this limitation by basing predictions on behavioural data obtained prior to the primary program executions; in contrast, dynamic prediction involves gathering and using such information during all executions. For the semi-static methods, the required information may be supplied by the programmer, or by the system software (e.g. the compiler), or obtained from "trial" runs with typical input data — a technique known as *profiling* — and then encoded as a single bit (appended to each branch instruction) indicating what the prediction is [42]. The semi-static methods are an improvement on the static ones, but in general they do not always work as well as the dynamic methods: for example, differences between the datasets used in profiling and those used in the final execution can affect the accuracy of semi-static methods.

4.4.2 Dynamic prediction

Dynamic prediction methods use run-time information on past branching behaviour to make the predictions. The most common implementation of such a strategy is based on a structure of the form shown in Figure 4.6 and is known as a *branch-target buffer* (BTB) or *branch-target cache* (BTC).[5] The main com-

---

[5] The reader should be aware that *branch target buffer* is used to refer to something quite different in old IBM machines.

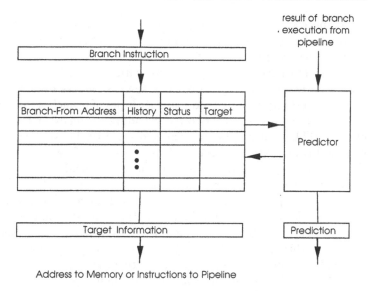

Figure 4.6.   Organization of dynamic branch predictor

ponent in that structure is a table that holds information on a number of the most recent dynamic branches in the program code; however, as we shall see below, not all the fields shown in the table need be explicit. The *branch-from address* (i.e. the address of the branch instruction) field (or some other information) uniquely identifies each branch instruction, the *history* field records some information on past behaviour, the *status* field holds information needed for the management of the table, and the *target* field holds such information as is needed to get the branch-target instructions into the pipeline. (We shall initially assume that the target-information is invariant relative to the branch-from address.) Condition codes may also be usefully be included in a BTB, as is done in the Motorola MC68060, for example, but this is rare.

The BTB works as follows. When a branch instruction is detected, its address is used to access the table, nominally by concurrent associative search on the contents of the Branch-From Address field[6]. If there is a match, then the Predictor uses the information in the History field of the selected line to make a prediction on whether or not the branch will be taken, and the instructions at the target are then fetched into the pipeline or not, depending on the prediction. And if there is no match, then a default prediction is made, with the target instructions being fetched or not, according to that prediction. In either case, when the direction of the branch has finally been resolved, a check is made to determine if the prediction was correct. If the prediction was incorrect, then

---

[6]A small table may be fully-associative and a large one set-associative: for example, the 256-line BTBs of the Intel Pentium and the Motorola MC68060 are 4-way set associative, and the 64-line BTB of the Hitachi Gmicro/500 is 2-way set-associative, whereas the 8-line one of the Manchester MU5 is fully-associative, as is the 32-line one in the Motorola 88110.

the effect of any (partially or wholly) processed instructions must be undone and the table updated. An example of an implementation of the basic scheme described above will be found in in Section 4.4.4.3.

Many implementations of a BTB store information only on conditional branch instructions, and we shall assume this to be the case unless otherwise stated. Nevertheless, a BTB may also be used for other types of branches (unconditionals and subroutine jumps) in order to reduce address latency or fetch latency or because the BTB implementation cannot distinguish between the different types of control-transfer (e.g. if the BTB is accessed before sufficient decoding has taken place). In the latter case, an additional field may be necessary to distinguish between the different types of instructions, as they will not all need the same type of prediction. Given that unconditional branches are always taken, the use of a BTB to speed up their execution can result in improved performance if straightforward code is manipulated to replace conditionals with unconditionals — this has been done with some success in the Gmicro/500 [115] — or if the BTB can simply assist in quickly getting the target instructions into the pipeline. We should also remark that although in many machines BTBs are used for dynamic branch prediction (based on changing history), and we shall generally assume this to be the case, it need not be so: the Motorola 88110, for example, uses a BTB but with static (op-code based) prediction [27].

The main issues in implementing a branch prediction unit of the type shown in Figure 4.6 are the size of the table, the contents of the various fields, the action to be taken when the table cannot be used, and when to make entries to and deletions from the table. Implementations vary in the details of these, and we next discuss some of the options.

4.4.2.1   The target information.
There are straightforward three choices for what to store in the Target field of the branch predictor. One option is to store the first few instructions at the target address; we shall refer to such a branch predictor as a *branch target-instruction buffer* (BTIB).[7] The second option is to store the address of the target instruction, which address is then used to prefetch the instructions if a "taken" prediction is made; in this case, we shall refer to the branch predictor a *branch target-address buffer* (BTAB). And the third option is to store the first few instructions at the target together with the first address of the subsequent instructions in that path and so have a hybrid of a BTAB and a BTIB. In terms of latency reduction, assuming correct prediction and prefetching of instructions, the address latency for a BTIB is essentially nil and that for a BTAB negative (if it is accessed early enough); the fetch latency is nil for both (assuming, again, that the BTAB is accessed early enough); and both buffers also reduce pipeline latency if they are accessed early enough. The BTIB may also be be usefully extended by the addition of a field to hold instructions on the alternate (straight-ahead) path and thus reduce

---

[7]A BTIB is also sometimes referred to as a *branch target cache*.

the delay in the event of an incorrect prediction; similarly, the BTAB may be extended to hold the alternate address.

Evidently, if the BTIB is to completely mask the branching delay, then the number of instructions in the line should be enough (in the case of an accurate prediction) to overcome the effects of the fetch and pipeline latencies. Depending on the memory access time and the number of pipeline stages before the control point, the number of instructions required may be too large to permit a practical implementation: for example, in the Manchester MU5 computer (for which a BTIB was initially considered), the shortest instruction is 16 bits wide and the ratio of the branch latency to the pipeline cycle time is 19, so BTIB would have required at least $19 \times 16 = 304$ bits in the target field of each line — a figure that was too large at the time the machine was designed and which is still quite large even today. In current machines, the ratio of branch-latency to pipeline-cycle is smaller than the 19:1 of the MU5 — typical figures range from 3:1 to 8:1 — and therefore a BTIB is more viable, especially when used to reduce just one of the three components of the branch latency. Thus, for example, the AMD Am29000 microprocessor uses a BTIB, with four instructions (a total of 128 bits) per line, the Motorola 88110 uses a BTIB with two instructions (64 bits) per line, and the GE RPM40 uses one with five instructions (90 bits) per line. It should also be noted that if the pipeline has a sufficiently large and readily-accessible (e.g. on-chip) instruction cache, then the instruction field of a BTIB need not exist separately if each line of the BTIB is directly associated with a line of the cache. Similarly, branch-to addresses in a BTAB may be just indices into the instruction cache, with full addresses being formed only as needed (e.g. on a cache miss).

Because the BTAB only holds addresses, all other things being equal, there are more constraints on its use if it is to give the same performance as a BTIB. In particular, it must appear much earlier in the pipeline, because the target instructions have to be obtained from memory; in other words, the reduction in address latency must be large enough to mask the fetch latency and (possibly) some part of the pipeline latency. Both types of buffer may be accessed after (partial) decoding, and therefore with only the address of branch instructions (and these are a small proportion of all instructions), but the BTAB is usually accessed before decoding takes place (i.e. instruction types are known), and in such a case it must be accessed with *all* addresses sent to memory. That is, the BTAB has to be accessed with *fetch-ahead addresses*, if no performance is to be lost, whereas the BTIB may be accessed with just the branch-from addresses. (In some implementations this problem of early access to the BTAB is solved by partially decoding all instructions before they enter the pipeline proper.) Consequently, to achieve the same performance, the BTAB must operate at average rate that is greater than that required of the BTIB or must appear much earlier in the pipeline, factors that, as we shall see below, have other consequences. Another implication is that it may be necessary to also use the BTAB for control-transfers other than conditional branches, even though it is only for these that a prediction is strictly required. Nevertheless, one can imagine variants of the BTAB that are designed for late placement: for example, we could arrange to always execute some instructions while the branch-target is being fetched; that is, to include delayed branching or branch-preparation (Section 4.3). As indicated above, a compromise between the BTAB and the BTIB is to have the target field store a few instructions at the target address and the starting address of the remaining instructions in that path. This hybrid, which is also a solution to the problem of a potentially long BTIB line, has been implemented in a few machines, such as the Hitachi Gmicro/200 microprocessor [60].

BTBs are most commonly used in branch prediction, and, unless otherwise specified, we shall assume that this is the case. Nevertheless, a BTIB (without history information) may be used solely to reduce fetch latency; in such a case it would be accessed only after the target address has been correctly determined. This type of implementation is most appropriate if the control point appears quite early in the pipeline. An example of such an implementation will be found in the AMD Am29000 (Section 4.4.6.2).

4.4.2.2   Stored addresses.

In a pipeline in which instruction-prefetching is used, there will be (at least) two registers that hold the addresses of instructions being fetched from memory or being processed: the program counter, which is updated for every instruction whose execution is considered complete, and the *prefetch counter*, which holds the fetch-ahead addresses used by the prefetching system. The prefetch counter will normally run ahead of the program counter, except when a branch is taken and its contents must be reset to match those of the program counter. Also, the prefetch counter must necessarily be at an early stage in the pipeline, whereas the program counter (or, more precisely, the point at which it is altered or at which branches are executed) may be at any stage: for example, in the Manchester MU5, the two are separated by five stages. Now, if the BTB is at an early stage — for example, if it is accessed before decoding — then it is the prefetching addresses that will be used to access it; that is, a branch is detected only when the fetch-ahead address matches a branch-from address.[8] But if the two counters are separated as indicated, then the branch-from addresses are not readily available to the prediction unit — at the time the branch instruction enters the pipeline, the address in the program counter will be that of some instruction that precedes the branch instruction — and these addresses must therefore be stored in the buffer. If, however, the branch-from addresses are available early enough and the mapping to the lines of the prediction table is simple enough to be carried out on-the-fly — for example, if part of the address is used to index the table, as in a direct-mapped cache, or if a simple hashing function is used — then the corresponding field

---

[8] An alternative is to use the program-counter values and then store in the BTAB the address of some instruction that precedes the branch instruction instead of storing the branch-from addresses. This requires that the instruction whose address is stored always be on the same execution path as the branch.

can be eliminated altogether. Other methods for reducing the width of the branch-from field are discussed in [40]. Note, however, that stored branch-from addresses also serve to uniquely identify each branch instruction for which there is an entry in the table.

It is also possible to reduce the width of the branch-to field, or to even completely eliminate it: If, say, the table is used only for branches that are encoded as displacements from the program counter (or some other register whose contents are readily available), then, evidently, only the displacements need be stored. A similar case is in a virtual memory system in which branches are encoded as displacements within a segment and, therefore, full addresses need not be stored. And another case is where part or all of the branch-to address is directly associated with a line in an instruction buffer or cache, and the size of the stored address can therefore be reduced or the address eliminated; the resulting system still resembles a BTIB. Thus, for example, in the Silicon Graphics TFP microprocessor, a 64-bit machine, the BTAB stores only the lowermost ten bits of the target address, with the remainder being obtained from the high-order bits of the predicted instruction-cache line [55]; the DEC Alpha 21164, the AMD-K5, and the Ultra-SPARC1 (all of which are discussed in detail below) are other examples of machines with this sort of implementation. Clearly, in such a case the details of the implementation must take into account the effect of cache-line misses and replacement. The branch-to addresses may also be completely eliminated from the table if they can be calculated quickly enough.

We shall refer to a table in which both the branch-from and branch-to addresses (or target information) have been removed as a *branch history table* (BHT); other terms in use are *branch prediction table*, and, for reasons that will become apparent below, *pattern history table*. With the branch-from address not stored, it is no longer possible to have a unique entry for each branch instruction, unless the table is as large as the address space, which is not practical. Therefore, in practice, different branch instructions will share an entry, with an simple accessing function of the type mentioned above. One important consequence of this arrangement is that there can be *history aliasing* between different branch instructions when one uses an entry that has been made by another, and this can affect performance; much of the current research in the design of branch predictors is devoted to finding ways to minimize such aliasing or its effects (Section 4.4.2.7). On the other hand, the lack of associativity means that larger tables (i.e. with more entries) can be realized at relatively little cost; consequently BHTs are typically much larger than BTBs of the type described above. A number of current machines employ a BHT, with (in some cases) additional structures. The MIPS R10000 is an example of a machine with such an implementation: the branch prediction table here consists of 512 lines, each of a 2-bit history field. When a branch instruction is processed, bits 11:3 of its address are used to index the table, thus doing away with the need for an explicit branch-from field; and the branch-to address is calculated by a dedicated 44-bit adder located at the same stage as the BHT [80, 121]. Similarly, the Mitsubishi Gmicro/100 uses a 256-line BHT, with one history bit per line [128]. The BHT may also be used in conjunction with an instruction cache, giving, in essence, a BTIB in which the branch-from address (or at least part of it) is implicit in the tag field of the cache. An essential difference between a BHT-cache combination and a BTIB is that the former will also store instructions other than branches. Furthermore, because an instruction cache typically stores several instructions in each line, the BHT-cache combination is more susceptible to thrashing if each BHT entry is associated with one line of the cache, as there may be cases where a cache line holds more than one branch instruction. Implementations of the BHT-cache combination will be found in a number of other machines, including the SPARC64, UltraSPARC-1, Mitsubishi Gmicro/100, and Silicon Graphics TFP microprocessors [55, 113, 119].

A BTB may also be designed to hold the addresses of both possible paths and so, by masking the address latency, reduce the delay after an incorrect prediction. Variants of this are implemented in the UltraSPARC and the Motorola MC68060 [21, 113]. Lastly, it may be better, in some cases, to store in the Branch-From field the address of some instruction that precedes the branch instruction itself; indeed, even the contents of the program counter may not necessarily be the branch-from address at the time the branch instruction is first encountered. Storing a different address can permit the fetching of target instructions to start earlier than would otherwise be the case — even before the branch instruction enters the pipeline — and, therefore, gives more scope in the placement of BTIB.

4.4.2.3   Variable target-addresses.

So far we have assumed that the target address is invariant. It is, however, possible to have the branch target change for an entry in the table, and this requires that the contents of the corresponding target-information field be changed accordingly: for example, in a return from a subroutine, the branch-from address is fixed but the branch-to address varies according to the point of call; this is also the case for some implementations of PASCAL-like CASE statements and similar constructs. Without some special action, such changes will lead to incorrect predictions: in one study over half of all incorrect predictions in some traces were found to be due to changes in the target address [63]; and in the Hitachi Gmicro/500, it has been found that also designing for subroutine-address changes reduces the branch penalty from five cycles to two [115]. Detecting the need for address changes in the table, and effecting these, is easiest with the BTAB: the detection can be accomplished by comparing the address at the control point (when the branch is resolved) with the address contained in the BTAB or by some other means. Alternatively, if only a few types of instruction can cause address change, then the need for this can be detected from the opcode, and a special structure (i.e. in addition to the BTB) may be used for such instructions. Another option is to simply take the the view that any BTB-entry that produces a number of consecutive mispredictions should be deleted, without determining precisely the cause of the mispredictions; this would then result in new entries being made without having to explicitly check for address-changes.

Although, to date most implementations of both BTAB and BTIB have not found the extra effort to be worthwhile and use hardware prediction only for invariant addresses, machines at the high end of the performance spectrum now do include some means to cope with changing addresses; some examples are the implementations in the DEC Alpha 21164 and Alpha 21264, the AMD K6, the Mitsubishi Gmicro/100, the MIPS R10000, the PowerPC 620, and the SPARC64. The essence of these implementations is to maintain one or more stacks that keep track of the subroutine jump and return addresses and then use the saved return addresses for the predicted targets. In one study, this has been shown to reduce mispredictions by about 20% [63]. Prediction mechanisms for indirect control-transfers are discussed further in Section 4.4.2.6.

4.4.2.4 Prediction table size and management. We have so far assumed that the prediction table holds an entry for every branch instruction that is encountered in the dynamic instruction stream. This is not likely to be possible because of the cost and performance implications of having a large, possibly associatively addressed, table. Thus the table will be of a finite size (and relatively small), and, as with an ordinary cache, a replacement algorithm must be used. Furthermore, the table size can be kept to a minimum if, say, it holds only entries for taken branches, with a default static strategy used for all instructions for which there are no entries. For replacement, assuming the BTB is not managed as a direct-mapped cache (as the BHT usually is), in which case the replacement algorithm is fixed, any of the standard algorithms used for conventional caches may be used. Algorithms that have been used to date include Not-Recently-Used (Manchester MU5), Random (AMD Am29000), First-In-First-Out (Motorola 88110), and Least-Recently-Used (AMD K6). In contrast with the usual situation for a data cache or an instruction cache, this is one situation where Random and First-In-First-Out replacement algorithms have been shown to be quite effective. Also, a genuine Least-Recently-Used algorithm need not be too expensive to implement here, as the number of lines in the typical table is usually quite small — as small 4 lines in the Hitachi Gmicro/200, 8 lines in the Manchester MU5, 16 lines in the AMD K6, and 64 lines in the NEC V80 and the Hitachi Gmicro/500. Nevertheless some recent machines have used much larger tables: for example, 256 lines in in the PowerPC 604 and Intel Pentium BTABs and 1024 lines in the DEC VAX 9000 BTAB.

Although in many implementations each line of the history corresponds to a unique line in the rest of the branch prediction table, this need not always be the case: if there are more history lines, then the loss of history whenever replacement occurs in the rest of the table can be avoided and better prediction achieved. Thus, for example, in the PowerPC 620, there are 2048 history lines but only 256 lines in the rest of the table. Such an arrangement evidently requires that the history field be accessed differently from the rest of the table; a common access method is to use part of the branch-from address to index the history, which means that several different instructions will now use the same History line. We may view this last arrangement as one of combining a large BHT with a small BTAB or BTIB; indeed, this is exactly what is done in machines such as the PowerPC 604, the PowerPC 620, the AMD-K6, and the HP-8000. One advantage of doing so is that by taking the BTAB or BTIB as one with an implicit single-bit history field, we essentially have three bits (in two "levels") for prediction and can, therefore, achieve greater accuracy overall. One research proposal goes farther and suggests the use of a multi-level prediction table in which the first level (of highest performance) is a BTIB, the next level is a BTAB, and the third level is a BHT [91].

4.4.2.5   Stored history.   There are two cases in which the prediction table, as so far described, will not be useful. The first case is when there is no entry in the table; this can easily be resolved by using a simple static or semi-static prediction method: for example, in such a case the VAX 9000 uses an opcode-based prediction, and the Manchester MU5 simply predicts that the branch will not be taken. (Although in principle any static or semi-static method could be used for the default prediction, the organization of the pipeline — in particular, the locations of the BTB and the control point — may rule out some options; this point is discussed further below.) The second case in which the prediction table is not useful is when there is an entry in the table, but it does not produce a correct prediction; this is usually due to insufficient stored history or, less often, to a change in the target-address.

Ideally, for the type of predictor described so far, the prediction for a given branch instruction will be based on the entire history of the execution of the instruction, but the potentially large amount of information that will have to be stored renders this impractical; therefore, in practice, the History field of the table consists of a fixed number of bits that are mapped onto a partial execution history. A common simple prediction is that the branch will be taken in the same direction as it was during its last execution and to predict that it will not be taken if it has not previously been executed. In this case, the history information in a line will consist of just one bit that records the direction of the last execution of the branch. If the table records only the most recent *taken* branches, then the history bit can be eliminated if a default prediction is used for branches for which there is no entry in the table; using the table for just the taken branches means that it will be smaller than would otherwise be the case. One-bit prediction is simple, and therefore inexpensive, but it does not take into account much of a branch's past behaviour: Consider, for example, the case when a branch is taken many times and then not taken, as in a loop-closure, and suppose that the loop-closing branch is for a loop that is contained in another loop. Then the next time the branch is taken the prediction (that it will not be taken) will be incorrect. Hence a single bit of history may not be very effective with nested loops. This type of prediction, in which a branch is simply categorized as "taken" or "not-taken", based on a single level of the history of its past executions, is known as *bimodal prediction*.

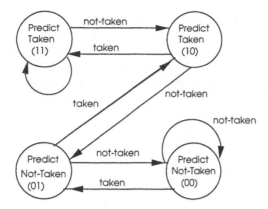

Figure 4.7.   Two-bit branch prediction

The one-bit prediction strategy just described can be improved upon by using two bits and changing the prediction only if the branch is not taken twice consecutively. (The two bits may be interpreted as *weakly taken, strongly taken, weakly not-taken*, and *strongly not-taken*, in contrast with just *taken* and *not-taken* of one-bit prediction.) This scheme, which is summarized by Figure 4.7, avoids the misprediction mentioned above for nested loops, as it requires two changes of state to change a prediction — a single loop exit will not necessarily change the prediction. A generalization would be to have a counter of arbitrary size, instead of just one or two bits, in the History field. This counter, a *saturating counter*, is then used as follows. It is initialised some threshold value (say, about halfway in the sequence of possible values); every time a branch is taken, the corresponding counter is incremented, unless its value has reached the maximum possible; every time the branch is not taken, the counter is decremented, unless its value is already zero; lastly, the prediction is made that the branch will be taken if the value of the counter greater than or equal to the threshold value and not taken otherwise. (If the counter values are equally divided between the *taken* and *not-taken* cases, then it is sufficient to examine only the most significant bit in order to make this decision.) In either case, entries may be made only for the most recent taken branches, with a default prediction for branches for which there is no entry. In two successive implementations (the Alpha 21064 and Alpha 21164) of the DEC Alpha architecture, for loops, the use of two bits of history was found to halve the number of mispredictions made when only one bit was used.

Although it would appear that the wider the counter the better, this is not necessarily so: A large counter requires more time to update than a small one (and may therefore increase the cycle time) and is also more likely to be more sensitive to the program run [102]. Based on studies that have been carried out so far, large counters offer little additional benefit and may even perform poorly when all else is taken into account. Some data on prediction accuracy relative to counter-size is given in Table 4.6, in which (a) corresponds to Tables 4.3 to 4.5; these show that 2-bit counters work well enough. Consequently, almost all machines built to date use just two bits for the counters. Some machines that use two-bit histories actually implement an algorithm that is slightly different from that of Figure 4.7. The alternative algorithm is shown in Figure 4.8; it differs from that of Figure 4.7 in the low-level details of what predictions are made for given patterns of directions, but it gives a similar performance.

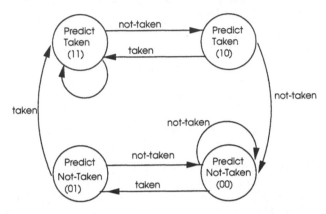

Figure 4.8.   Alternative two-bit branch prediction

4.4.2.6   Two-level adaptive prediction.

A recent development in hardware branch prediction is that of *correlation-based* (or *two-level adaptive*) *branch prediction*, in which the prediction made for a given branch is dependent on the histories of related branches [44, 90, 98, 122, 124]. The rationale for this approach is that whereas in loop-closing branches the past history of an individual branch instruction is usually a good guide to its future behaviour, with more complex control-flow structures, such sequences of IF-ELSE constructs or nestings of similar constructs, the direction of a branch is frequently affected by the directions taken by related branches — for example, by preceding branches. If we consider each of the possible paths that lead to a given nested branch, then it is clear that prediction in such a case should be based on the subhistories determined by such paths, i.e. on how a particular branch is arrived at, rather than on just the individual history of a branch instruction. And in sequences of conditionals, there will be instances when the outcome of one condition-test depends on that of a preceding condition if the conditions are related in some way — for example, if part of the conditions are common, if part of one is a complement of the other, if one condition is a logical combination of others, and so forth. Accordingly, in correlation-based branch prediction, each prediction takes into account the histories of several branch instructions executed in the immediate past.

The branch predictor of Figure 4.6 may modified into a two-level adaptive predictor (Figure 4.9) by making the following changes: A k-bit shift register, the *branch-history register* — k is known as the *history depth* — is added to record the directions taken by the k last branch instructions; and as each

Table 4.6.   Prediction accuracy (%) vs. counter size

(a)

n	IBM System/370				DEC	CDC
	CPL	BUS	SCI	SUP	PDP-11	6400
0	64.1	64.4	70.4	54.0	73.8	77.8
1	91.9	95.2	86.6	79.7	97.5	82.3
2	93.3	96.5	90.8	83.4	97.5	90.2
3	93.7	96.6	91.0	83.5	97.1	93.4
4	94.5	96.8	91.8	83.7	98.7	94.8
5	94.7	97.0	92.0	83.9	98.2	95.1

(b)

n	bf COMP	TEXT	FP	SPARC	VAX	68K	Avg.
0	67.21	66.13	61.78	56.80	68.85	64.96	64.29
1	86.30	94.84	85.03	87.51	89.71	85.55	88.15
2	87.81	95.45	86.72	88.69	91.56	86.27	89.42
4	88.66	96.02	87.85	90.61	93.36	88.21	90.79
8	89.19	96.18	89.28	91.78	93.70	89.00	91.52
16	90.14	96.40	89.90	92.93	95.25	91.48	92.68

MIPS=traces from MIPS assemblers, compilers, etc.
FP=traces from MIPS floating-point programs
TEXT=traces from MIPS text processing programs
SPARC/VAX/68K=traces from various programs on VAX/SPARC/M68K

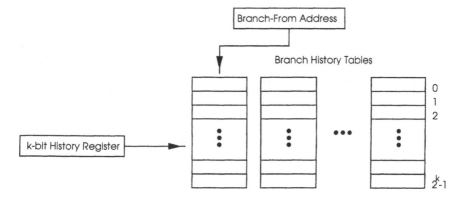

Figure 4.9.   History access-table in correlated-branch predictor

branch instruction is executed, the register is shifted left and a 1 or 0 entered at the right-hand end, according to whether the branch was taken or not. The single history subtable of Figure 4.6 is now replaced by several tables — one for each (part of the) branch-from address, with each table having 2^k entries for the different the subhistories that may be recorded in the branch-history register. The history to be used in making a prediction for a given branch is now accessed by using the branch-from address (or part of it) to to select one of the history tables and then using the the contents of the branch-history register to select an entry within that table. This arrangement therefore ensures that the prediction made for each branch is a function of the pattern of directions or history taken by the most recent k related (by address) branches. Thus there are two levels of history used to arrive at a prediction: the k bits indicate the amount of global history taken into account, and the m bits indicate the amount of local history taken into account, i.e. the particular history for the *pattern* represented by the k bits. (The BHT in this context is also frequently referred to as a *pattern history table*.) A predictor that uses k history-register bits and m-bit counters in the history table is known as a (k, m)-*predictor*; so the simple predictors previously described are $(0, 1)$ and $(0, 2)$ predictors.

For path-based prediction, the branch history register is replaced by a register that records *path history* instead of pattern history, where the path history is constructed by concatenating — by shifting into the path register — some bits from each of the branch-from addresses of the most recent branches or by hashing together such addresses [87]. Concatenating only parts of addresses means that there will sometimes be a failure to distinguish between two different paths, leading to aliasing in the use of table entry; however, concatenating the result of concatenating full-length addresses, but as the result could be rather long. A path-based prediction system in which the path-length is variable has been proposed and shown to yield very good performance [107].

There are a number of other possible variants on the basic two-level predictor as described above. For the history register, we have three main options: One is to have a single branch-history register, in which case the prediction for a given branch is (potentially) affected by the directions taken by all branch instructions in the dynamic stream (as these share the single register); in this case the predictor is known as a *global adaptive* (*GA*) *predictor*. A second option is to have one history register for each branch-from address (i.e. branch instruction), in which case the prediction for a given branch is determined purely locally — by just recent patterns of execution of the same instructions; in this case the predictor is known as a *per-address adaptive* (*PA*) *predictor* or *local predictor*. And in between these two extremes, we may one history register for each *set* of branch instructions, where membership in the set is determined by address (e.g two instructions are placed in the same set if they are in the same instruction block), by opcode, and so forth. In this case the prediction for a given branch is affected by the directions taken by branches in the same set, and the predictor is known as a *per-set adaptive* (*SA*) *predictor*. So, for a (k, m) predictor, if the predictor is GA, then the k bits will record the directions taken by the most recent k branch instructions; if the predictor is PA, then the bits record the directions taken taken in the last k executions of the same branch instruction; and if the predictor is PA, then the bits record the directions taken by the most recent k branch instructions in the same set.

Further, for each of the three broad classes above, there are three possible arrangements for the BHT: one global (g) table for all branches, one table for each set (s) of branches, and one table for each branch-from address (p). We therefore end up with a total of nine possible organizations: GAg, GAs, GAp, SAg, SAs, SAp, PAg, PAs, and PAp. These have been compared in detail, with the following conclusions [124]: The GA schemes are the best performers on integer programs; such programs tend to have many IF-ELSE sequences that are highly correlated and for which global prediction is very effective. But the GA schemes, because they tend to map different branches onto the same BHT-entry, require longer branch history registers or more history-table entries if they are to be most effective; they are also more expensive to implement than the PA schemes. The PA schemes are the best performers on floating-point programs; these tend to have many loop-closing branches, and the PA schemes are very effective on such localized behaviour. The PA schemes also require fewer history-table entries (because there is less aliasing) and generally cost less than either GA or SA schemes. The SA schemes, as we might expect, have a performance (with respect to type of program) that is between those of the GA and PA schemes but, as a consequence of the required set-partitioning, are also the most expensive to implement. Overall, the study from which these results are summarized has shown that if cost not an issue, then GAs is the best predictor; otherwise, PAs is the best.

The accessing of the history tables in Figure 4.9 may also be viewed (and frequently is implemented) as one on a one-dimensional table that is accessed via an index obtained by concatenating (part of) the branch-from address with the contents of the branch-history register. That is, part of the index selects a section of the table, and the other part determines a displacement into that section. (Unless otherwise stated, we shall henceforth adopt this view of the multiple tables as a single table.) So a straightforward way to implement GAs is to concatenate low-order bits of the branch-from address (i.e. place instructions in the same set if they have the same high-order address bits) with the contents of the branch-history register and then use the result to index the BHT. The resulting predictor is known as *Gselect*, because it uses selected bits of the branch-from address. (Note that, in general, the selected could also have been from the middle or high-order bits of the branch-from address.) An implementation of Gselect is in the AMD-K6, in which 9 bits of history register are combined with 4 bits of program counter to yield a 13-bit index into an 8192×2-bit BHT. An alternative method of combining the two bit sequences is to use an exclusive-OR operation; such a predictor, known as *Gshare*, is described below and has been shown to have better performance that Gselect. Other combinations may be obtained by interleaving the bits, hashing, and so forth.

There are a number of useful modifications and extensions than can be made to the basic two-pevel predictor, many of which are discussed below. For example, one is to modify the use of the predictor so that indirect branches (i.e. variable target addresses) are handled with fewer mispredictions than would otherwise be the case. One proposal for this suggests the use of a *target cache* that holds the target addresses of indirect branches. To make a prediction, the branch-from address and pattern-history or path-history and combined and then used to access the target cache for the destination address; and after the branch instruction has been executed, the result is used to update the target cache. Any of the methods suggested above may be used to combine the two addressing bit-sequences, and the cache itself may be implemented with or without tags. Simulations indicate that XOR is better than concatenation, that which of pattern-history or path-history gives the better performance may be program-dependent, and that a tagless cache is better than a tagged one with a low degree of associativity. Related work that explores this design space in greater depth and breadth will be found in [30].

Results obtained from simulations and implementations of two-level adaptive predictors have been quite good: for example, on the SPECint benchmarks, using a BTB with an 8-bit history register and a total total history-table size of 1Kbytes has been shown to yield a performance improvement of up to 11% over a BTB that has a history table of the same size but no branch-correlation [90]. Other results from the comparison of a basic bimodal BTB and a correlated-branch BTB shows that for small buffer sizes the basic BTB performs better, and for large sizes the correlated-branch BTB is better [39]. This last result is consistent with the fact that the basic predictor keeps a unique history for each branch instruction, whereas the two-level predictor (unless the history table is quite large) maps several branches into the same entry of the history table and so suffers from interference. Several of the latest machines (e.g. the Intel Pentium Pro and the DEC Alpha 21264, which is described below) implement some form of correlated branch prediction. A static version of the hardware correlated-branch prediction method has also been developed. This involves making extra copies (at compile-time) of code along paths that have correlated branches and then encoding the branch prediction as extra bits to the program counter [44, 130]. A recent analysis of why two-level predictors work (and where they fail) will be found in [38].

4.4.2.7   Reducing aliasing in BHT.   Because the basic BHT is accessed in the same way as a direct-mapped cache (one without tags, though), different branches can be mapped onto the same entry, and so predictions for one branch will sometimes be made using values created by a different branch. This is more likely to be the case for global than for local schemes and can affect the prediction accuracy. The effect of such *aliasing* (or *interference*) may be classified as *positive* (or *constructive*), if it improves accuracy, *negative* (or *destructive*), if it reduces accuracy, and *neutral* (or *harmless*), if it has no effect. Detailed studies of aliasing and its effects will be found in [99, 109]. A number of schemes have been devised to deal with aliasing, and we next look at some of these.

The first, and best known, of the schemes that try to reduce negative aliasing or its effects is *Gshare*, essentially a variant of GAs that resembles GAg. Gshare is based on the fact that global branch histories are not uniformly distributed, and therefore not all all combinations of of the contents of the branch-history register and the branch-from address (as used in Gselect, say) are useful: frequent global branch-history patterns tend to be sparse and therefore simple combinations, e.g. concatenation, of the two indices encode some redundant information [75]. As a consequence, hashing the two indices into one does not necessarily lead to a loss of information and can improve the prediction accuracy, by allowing the use of more bits of both global history and branch-from address or by separating branches that would otherwise map onto the same entry. Thus whereas accessing an n-entry table requires that the concatenation of the two indices in Gselect be $\log_2 n$ bits wide, with hashing *each* of the two indices may be up to $\log_2 n$ wide. One method of hashing the two indices that has been shown to be effective is to use an exclusive-OR operation (Figure 4.10). (If fewer bits of global history than of branch-from address are used, then the XOR operation is on the high-order bits of the latter, as these tend to be more sparse than the low-order bits.) Gshare has been shown to improve accuracy at little additional cost and currently appears to be more favoured than other types of predictor.

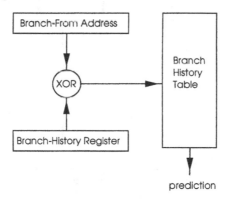

Figure 4.10.   Gshare branch predictor

The essence of most of the conflicts in the basic predictor is the lack of associativity. One could try to deal with this by implementing the BHT as a set-associative or fully-associative structure, but this might not be very cost-effective, given that the typical address tag would be much larger than the one or two bits in the data (counter) field. The *gskewed predictor* is designed on the basis that multiple banks of only counters can, if used appropriately, effectively emulate higher associativity [81]. Thus in a Gskewed predictor, the single BHT above is replaced with several banks, all of which are accessed for each branch — but with a different accessing (hashing) function for each bank (Figure 4.11). Each bank yields a preliminary prediction, and a majority vote is then used to determine the final prediction. The rationale for this arrangement is that by using different accessing functions, the positions selected within the different banks are skewed relative to one another, and therefore two branches that map onto the same entry in a single bank are unlikely to map onto exactly the same entries within the multiple banks; so aliasing in one bank has little effect if the prediction is determined by the other (presumably) no-aliased banks. After a branch instruction has been executed, only the counters in those banks that gave the correct prediction are updated; this turns out to give better performance than also updating banks that give an incorrect prediction. Performance studies have shown that a Gskewed predictor with $3n$ BHT entries (three banks of n entries each) gives about the same performance as a fully-associative BHT with n entries and Least Recently Used replacement and, when all other factors are taken into account, outperforms Gshare.

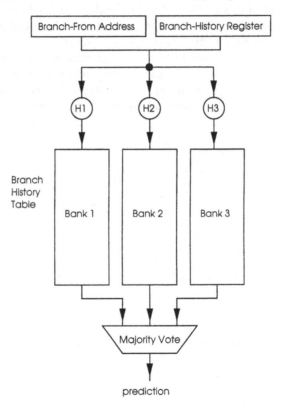

Figure 4.11.   Gskewed branch predictor

Gshare basically tries to reduce all negative aliasing. A different approach, used in the *agree predictor*, is to try and convert negative aliasing into positive or neutral aliasing. The basic idea here is to take advantage of the fact that some branches tend to go in one direction most of the time, i.e. they are *biased*, and are therefore easy to predict. The Agree predictor uses a BHT-entry in a slightly different manner from what has been assumed so far: instead of using the entry to record the directions taken by a branch, it is used to record the extent to which the directions taken agree with what is considered to be the most likely direction to be taken (i.e. the direction in which the branch is biased). Thus a *biasing bit* is associated (either dynamically, e.g. via a BTB of instruction cache, or statically) with each branch instruction, the BHT-counter is incremented if the corresponding branch goes in the direction indicated by biasing bit and decremented otherwise, and a prediction is made according to the agreement (Figure 4.12). The essential idea is that if two branches use the same BHT-entry, then, provided the biasing bit is set appropriately, both branches will most likely update the counter in the same direction, thus converting negative aliasing into positive aliasing (or at least neutral aliasing). Nevertheless, the scheme is not perfect: it does not work in those instances where a branch is biased but in the direction opposite to that indicated by the bias bit. Negative interference can then still occur between those branches that conform to the biasing bit and those that do not. Note that the Agree operation only affects the use of the BHT, and the method may therefore be used other types of predictor. Experiments that combine Agree with Gshare have shown that, relative to other methods, it is very effective with small BHTs but less so with large ones; this is as we should expect, given that larger BHTs have less aliasing.

The HP-8500 and the Centaur C6+ are two examples of machines that implement Agree predictors. In the HP PA-8500 the basic prediction is static, and the bias bit is just the static-prediction bit that is appended to each branch instruction. The Centaur C6+ also uses static prediction, but in this case the static-prediction (bias) bit is formed on-the-fly, by predicting according to the direction of the branch; the BHT itself is a 4096×1-bit table [49].

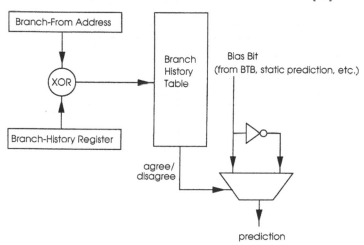

Figure 4.12.   Agree branch predictor

The *bi-mode predictor* also tries to turn negative aliasing into at least neutral aliasing [68]. In this case the single BHT is replaced with three tables (of 2-bit saturating counters): one BHT that keeps track of "taken" directions, one BHT that keeps track of "not-taken" directions, and, for each branch, one "choice" table that chooses which of the two to use for the prediction (Figure 4.13). The branch-from address and the contents of the history register are combined as in Gshare and then used to access the direction-BHTs, but the choice-BHT is accessed with just the branch-from address. Once the branch instruction has been executed, the direction-BHT from which the prediction was taken is updated. The choice-BHT is updated only if makes a choice that agrees with the branch execution outcome or if the chosen direction-BHT gives an incorrect prediction.

The essential idea in the Bi-Mode predictor is that because negative aliasing occurs when two branch instructions share a BHT-entry but are taken differently, it can be reduced if the branch instructions are grouped according to likely outcome: branches that are likely to be taken will have their predictions in the "taken"-BHT and those that are likely not to be taken will have their predictions in the "not-taken"-BHT, thus reducing the probability that a taken branch will share a BHT-entry with a not-taken one. The predictor therefore reduces negative aliasing by making predictions according to the biasing of the branch directions, but unlike the Agree predictor, which also does the same thing, here the biasing is dynamic — the bias information is in the table used to make the final predictions. Nevertheless, dynamic adjustment of the bias does not eliminate aliasing between those branches that conform to the bias bit and those that do not. Results from simulations have shown that the Bi-Mode predictor generally gives better performance than Gshare but is cheaper to implement.

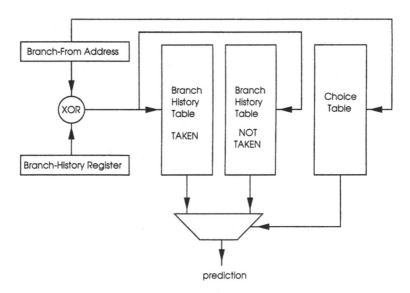

Figure 4.13.   Bi-Mode branch predictor

Evidently, the structure of the Bi-Mode predictor may be generalized to similar arrangements, of a *selective predictor* (or *hybrid predictor*), in which the two BHTs serve different functions [75]. One combination, suggested by the general discussion above of the GA, PA, and SA schemes, is to use one BHT for local predictions and one for global predictions (Figure 4.14) and so have good performance on both integer and floating-point programs. Thus one predictor, P1, makes a local prediction; another predictor makes a global prediction; and a Choice table of 2-bit counters keeps track (in the manner shown in Table 4.7) of which of the two predictors is more accurate and makes a final prediction accordingly. Various such combinations have been studied, and it has been shown that Local/Gshare and Bimodal/Gshare (cf. PAs/GAs), for example, give very good performance, with the former having the edge when the history tables are large [75]. Such a combination is currently implemented in the DEC Alpha 21264 (Section 4.4.4.6). The arrangement can be extended to more than two predictors (although only two have been used to date) and according to other classifications of branch instructions: one possibility is to divide (at compile or during profiling) branch instructions into different groups, according to the probability with which they are taken, and then use a different predictor for each group [15]. The different preliminary predictions combined may include static, semi-static, and dynamic predictions, and the final prediction used for a given branch may be chosen at compile-time, after profiling, or during execution. Experiments with such combinations have shown that the addition of profile-guided semi-static prediction to the type of predictor shown in Figure 4.14 gives a non-trivial improvement in performance.

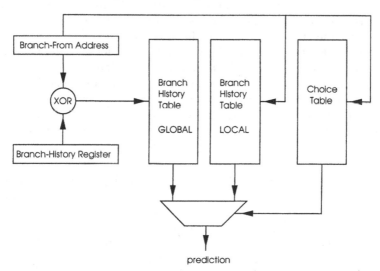

Figure 4.14.   Selective branch predictor

The *filter predictor* is based on the observations that ordinarily there is some redundant information stored in the BHT and that one bit is sufficient to predict

Table 4.7.   Choice-table update in selective predictor

P1	P2	Choice
0	0	no change
0	1	decrement counter
1	0	increment counter
1	1	no change

0=incorrect, 1=correct

the directions of highly biased branches [18]. The goal, then, in the design of the predictor is to filter out the highly biased branches and store in the BHT only information for the remaining branches, thus reducing all types of aliasing. The arrangement used for this is shown in Figure 4.15. A bias bit is employed in the BTB (or elsewhere), as in the Agree predictor, and a saturating counter is used in the BTB to keep track of the strength of the bias. Whan an entry is first made in the BTB, the bias is set (to the direction taken in the first execution or from a static/semi-static prediction), and the counter is initialized (to its maximum value). Subsequently, whenever a branch instruction is executed, its BTB-counter is incremented if the direction is the same as that indicated by the bias bit; otherwise the counter is returned to zero, and the bias bit is complemented. The prediction for each branch is made from the bias bit or from the BHT counter: If the counter is saturated, then it means the branch is strongly likely to be taken in the direction indicated by the bias bit, and so the bias bit is used for the prediction; in this case the BHT-entry is not updated. Otherwise, the BHT-entry is used to make the prediction and then updated according to the outcome of the branch execution. The filtering occurs at the point at which highly biased branches are stopped from updating the BHT, and its effect is to reduce all types of aliasing. As with other predictors that use similar mechanisms, the predictions here will be incorrect in those instances where the direction taken by a highly biased branch is not that indicated by the bias bit.

The latest two-level predictor that tries to reduce aliasing and its effects is the *YAGS (Yet Another Global Scheme)* predictor [33]. This predictor tries to combine the best features of some of those described above but without any of their drawbacks: As in the Agree and Bimode predictors, YAGS partitions branches into two groups, according to the bias that they will most likely be taken or most likely not be taken; and like the Filter predictor, YAGS seeks to minimize the amount of useless information stored in the BHT. However, unlike these other predictors, the design of YAGS takes into account those instances in which the direction taken by a branch is different from that indicated by the bias, and, accordingly, tries to reduce mispredictions in these cases.

The design of YAGS is shown in Figure 4.16. The choice-BHT is similar to that in the Bomode predictor and stores a dynamically adjusted bias. The

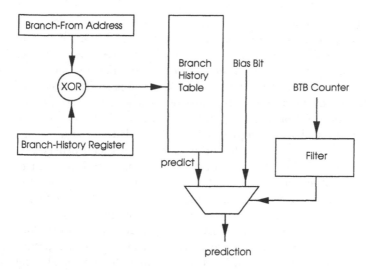

Figure 4.15.   Filter branch predictor

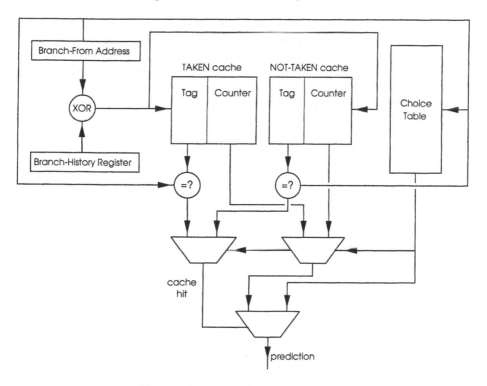

Figure 4.16.   YAGS branch predictor

Taken and Not-Taken caches are similar to direction BHTs in the Bimode predictor but serve a slightly different role: they keep track of cases in which a branch-direction does not conform to the indicated bias. Each entry in one of these caches consists of the usual 2-bit counter and a small tag that consists of the least significant bits of the branch-from addres; the tags distinguish between branch instructions sufficiently to eliminate aliasing between consecutive branches. For each branch, the first access made is to the choice-BHT. This is then followed by an access to the direction-cache that corresponds to the opposite direction to that indicated in the choice-BHT; this check is to determine if the prediction does not agree with the bias. If there is a hit in the chosen direction-cache, then the prediction is made from there; otherwise, it is made from the choice-BHT. The updating of the choice-BHT is as in the Bimode predictor; a direction cache is updated only when it is used for the final prediction or if the choice-BHT is used for the prediction but the final outcome of the branch execution does not agree with that prediction.

Performance simulations on SPEC95 benchmarks have shown that the YAGS predictor (with 6 to 8-bit tags) performs better than other predictors if the BHTs are small; for large BHTS, the natural reduction in aliasing gives the YAGS predictor a much smaller edge.

4.4.2.8 Three-level adaptive branch predictors.

A few studies have shown that for two-level prediction, the best predictor is one that changes history length according to the particular type of program run: GAs, for example, does not work equally well on both mixed-direction branches and highly-biased branches but performs better when two global-branch-history lengths are used — a long one for mixed-direction branches and a short one for biased branches [15]. (Although a long history ought to be better, in general, because it captures more of the program behaviour, it also implies a larger BHT (relative to fixed number of branch-from bits) and a longer time to initialize; the latter means that the BHT cannot quickly adapt when program execution changes from one phase to another.) Accordingly, one recent study proposes the use of a third level of adaptivity: dynamic selection of the history length [62]. This is known as *dynamic history-length fitting* (DHLF) and may be used in conjunction with any of the two-level predictors described above. As an example, Figure 4.17 (lower half) shows the generation of the BHT-index (of 0 to 6 bits) in a DHLF-Gshare predictor, relative to plain Gshare (upper half of Figure 4.17); the number of branch-history bits included in the XOR operation is variable, from zero to the size of the branch-history register. The main components in the additional DHLF unit are: a *misprediction table* of as many entries are the number of bits used to index the BHT and in which entry k holds the number of mispredictions that have occurred in the last execution interval that a history of length k was used, a pointer to the entry in the misprediction table that corresponds to the number of history bits currently in use, a pointer to the table entry that contains the smallest misprediction count, a counter that keeps track of the number of mispredictions in the current interval, and a counter that keeps track of the number of predicted branches in the current interval.

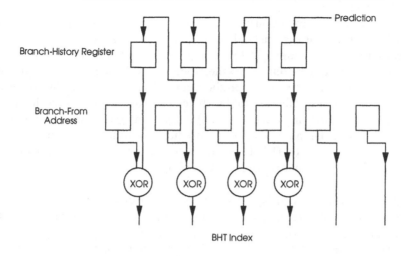

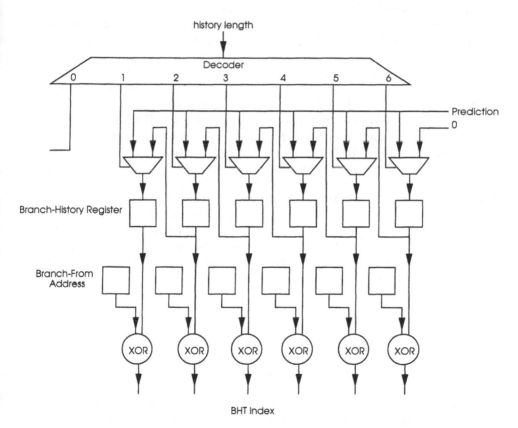

Figure 4.17.   Dynamic history-length fitting

At the start of an execution run, the two counters, the number of history bits, and all the entries in the misprediction table are all set to zero. Subsequently, the various values are appropriately changed, and an interval is considered complete when the branch-counter reaches some pre-specified value. At the end of an interval, the value in the misprediction counter is stored in the corresponding entry in the misprediction table, the current history length is changed (by incrementing or decrementing it by one) if the number of mispredictions exceed the minimum in the table, and the misprediction and branch counters are then reset. Simulations of the DHFL-Ghsare predictor has shown that it can achieve near-optimum performance across a variety of program types.

4.4.2.9   Remarks on mispredictions. The recovery time and costs in the event of a misprediction is not just the time to calculate the correct target address and fetch the correct instructions: it also includes the time required to undo the effects of any partially or wholly processed instructions. The latter part of this *misprediction penalty* is determined by a number of factors, including the instruction-set architecture, the particular pipeline implementation, and the degree to which instructions are processed before recovery is initiated. As a simple example, in the Manchester MU5 instructions are speculatively processed only up to the control point, which is about halfway in the pipeline (and before the execution units), but the instruction set allows for stack operations that are carried out before the control point; this necessitates the use of extra registers to hold various stack pointer values in case a rollback is required (see Figure 1.8). In some current high-performance microprocessors, speculative processing is carried right through to the execution stage, and so the amount of state to be saved and restored can be quite large. There are, moreover, additional complications when instructions are executed out-of-order; these include determining what instructions logically preceded the branch and what instructions logically came after it then becomes more difficult. Machines such as the PowerPC implementations use a relatively complex system of tags to determine the proper order in which partially processed instructions should be undone, and others employ even more complex mechanisms. The general problem of how to precisely restore machine-state in the event of an exception is discussed in detail in Chapter 7.

4.4.3   Case studies

We now look at a number of practical implementations of instruction-buffering with branch prediction or the use of BTBs. The first is the Motorola 68040 implementation and is an example of simple static prediction. The second is the implementation in the Advanced Micro Devices Am29000 and is an example of a BTIB that is used for the reduction of just the fetch latency, with no prediction involved; it also shows the kind of complications that can arise in the implementation of even a simple BTB [3]. The third example is the BTAB used in the Manchester MU5; this is an example of a BTAB that also shows how technology and architecture (instruction set) can affect implementation

[84]. The fourth example example is the system used in the PowerPC 604 and PowerPC 620; it is an example of the state-of-the-art and employs two types of branch predictor, consisting of a BTAB and a BHT [28, 105]. The fifth example is the implementation of the SPARC64, another state-of-the-art microprocessor; this shows the use of a BHT with an instruction cache and a return-address stack to handle variable jump-to addresses [119]. The next pair of examples are the branch prediction systems used in the DEC Alpha 21164 and Alpha 21264; both are additional examples of branch prediction in a state-of-the-art microprocessor, also with the capability to handle variable branch-to addresses [34, 50]. The Alpha 21264 also employs a selective predictor that combines global/local prediction, which makes it one of the most sophisticated systems currently implemented. The last two examples are the systems used in the AMD K5 and K6; both systems use a combination of various methods discussed above [4, 20, 52].

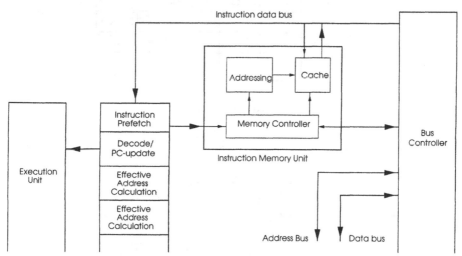

Figure 4.18.   Instruction-fetch unit of in Motorola 68040

4.4.3.1   Motorola 68040 instruction fetching.   The first two stages of the instruction pipeline and the instruction-fetch unit of the Motorola 68040 are shown in Figure 4.18. The first stage of the pipeline is the Instruction Prefetching stage, which generates instruction addresses for the Instruction Memory Unit. This stage contains an 8-word×16-bit buffer to hold prefetched instructions and is designed to supply instructions to the rest of the pipeline at a rate of three 16-bit instructions per cycle.

The Motorola 68040 handles branches by using the simple prediction that all branches will be taken. Thus for all branches, the control adder (in the second stage of the pipeline) adjusts the value of the program counter before complete decoding takes place, and requests for the addressed instructions are sent to memory. In the case of a conditional branch, to guard against the possibility of an incorrect prediction, when a change of flow takes place, the alternate value of the program counter, the instruction in the third stage of the pipeline, and the contents of the instruction buffer are all saved in special registers. (These extra registers are an example of the the hardware costs of squashing partially processed instructions.) If, after the evaluation of the branch condition, it turns out that the branch should not have been taken, then a roll-back is carried out to restore the saved information.

Stage

		IP	D/PC	EA	FE	EX
	k	Branch	Test			
	k+1	Next	Branch	Test		
	k+2	Target	Next	Branch	Test	
Cycle	k+3		Target		Branch	Test
	k+4			Target		Branch
	k+5				Target	
	k+6					Target

(a) *taken* branch

Stage

		IP	D/PC	EA	FE	EX
	k	Branch	Test			
	k+1	Next	Branch	Test		
	k+2	Target	Next	Branch	Test	
Cycle	k+3		Target		Branch	Test
	k+4		Restore	Restore		Branch
	k+5			Next		
	k+6				Next	
	k+7					Next

(b) *not-taken* branch

Figure 4.19.   Branch-instruction flow in Motorola 68040

Figure 4.19 shows the pipeline flow for both *taken* and *not-taken* branches. The figure shows the performance-reducing "bubbles" (the shaded areas) that occur in the pipeline for both types of branches, with the not-taken case being the worse: For the taken branch, the instruction in the straight-ahead path is fetched into the pipeline while the branch is being decoded and is then discarded after being decoded, with the branch-target instruction proceeding through the pipeline. The non-taken case starts off similarly, but the target instruction is discarded after being decoded and a restore operation that requires an extra cycle is initiated to bring back the instructions on the straight-ahead path.

4.4.3.2 AMD Am29000 instruction fetch unit. The AMD Am29000 pipeline consists of four main stages: FETCH, DECODE, EXECUTE, and WRITE-BACK. In the first stage, the Instruction Fetch Unit (IFU) tries to keep the pipeline busy by supplying one instruction in each cycle. The main components of the IFU are shown in Figure 4.20; these are the Instruction Prefetch Buffer, the Branch Target Instruction Buffer, and the Program Counter Unit. The BTIB is accessed with target addresses in the program counter and for branches that are definitely taken. There is therefore no prediction, and the main goal is to reduce just the fetch latency.

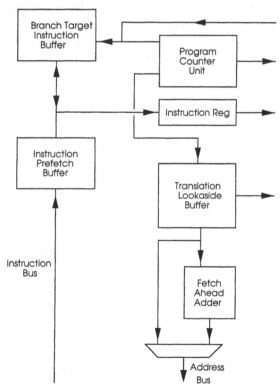

Figure 4.20.   Am29000 instruction fetch unit

During normal sequencing, lookahead addresses, computed by the Fetch-Ahead Adder and stored in the Instruction Fetch Pointer, are sent to the main memory unit and are used to prefetch instructions, which are then stored in the Instruction Prefetch Buffer. The buffer is a four-word queue that the IFU, by fetching instructions at least four cycles before they are required, attempts to keep full at all times,.

Branches are handled by using software delays and the BTIB. The BTIB is two-way set-associative, with each set consisting of 16 four-word (four-instruction) lines. Whenever a non-sequential instruction-fetch is required, the target address is sent to both the BTIB and the memory management unit; the latter immediately translates the address. If the target instruction is in the BTIB, then it is read out and in the next cycle is sent to second stage of the pipeline (i.e. the Decode stage). (During the subsequent cycles the other instructions in the same line are also sent to the Decode stage.) At the same time, the fetch-ahead logic also initiates instruction-fetching four words beyond the target address. If the target is not in the BTIB, then the memory management unit uses the translated address to initiate instruction-fetch from memory; the first four instructions at the target address are sent to the Decode stage as they arrive and also loaded into a randomly selected line of the BTIB. (According to the designers, in this case Random replacement gives slightly better performance and is cheaper to implement than Least-Recently-Used.)

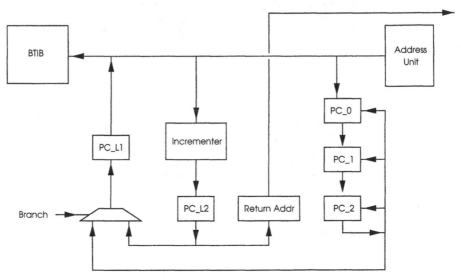

Figure 4.21.   Am29000 program counter unit

There are two cases that require special handling in the operation of the BTIB: The first is when a virtual-page address-boundary is crossed while loading a line of the BTIB. Prefetching stops in such a case, and the BTIB line will not be full. If the following instructions are required, then they are fetched as needed and loaded into a randomly selected line of the cache. The second special case is when one of the first four instructions the target stream is itself a branch instruction. If this is either of the first two instructions, then that branch will be taken before the selected cache line is filled, and the line will contain less than the usual four instructions. If the branch is successful, then the fact that the line is only partially filled does not affect performance, because any instructions after the branch's delay-instruction in the line would not be required anyway. If, however, the branch is unsuccessful, then the instructions after the branch must be fetched and loaded into the same line, which then has validity bits set.

The Program Counter Unit forms and sequences addresses for the IFU and also assists with the handling of interrupts and traps. It has the organization shown in Figure 4.21. The counter is implemented as a master-slave pair of registers: the master holds the address of the current instruction and the slave holds the next sequential address. The Return Address register holds the address of the instruction that follows a subroutine call. The three registers PC0, PC1, and PC2 hold the addresses of the instructions in the last three stages of the pipeline. When a return from an interrupt takes place, the contents of PC0 and PC1 are used to restart the pipeline. Two registers are required for the restart because of the use of delayed branching: an interrupt can occur while the instruction in the delay slot is being executed and the target instruction is being decoded; in such a case PC1 holds the address of the delay instruction, and PC0 holds the address of the branch target. PC2 is used to hold the addresses of instructions that cause other exceptions.

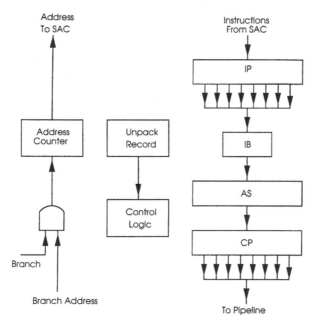

Figure 4.22.   MU5 instruction buffer unit

4.4.3.3 MU5 instruction buffer unit and branch target buffer. The fetching and buffering of instructions in the MU5 instruction pipeline (see Section 1.5.3) is carried out under the control of the Instruction Buffer Unit and a

BTAB. The Instruction Buffer Unit has the organization shown in Figure 4.22. The performance goal of the unit is to supply the pipeline with one 16-bit instruction *parcel* at every cycle. Under normal sequencing, the Address Counter generates sequential addresses (that differ by 2) in the Advance Counter (AC) and sends these to the Store Access Control Unit (SAC). The SAC returns 128-bit words (two 64-bit memory words) that are then buffered in four registers — Input (IP), Intermediate Buffer (IB), Additional Storage (AS), and Close Pack (CP) — before being forwarded to the rest of the pipeline. The three registers IP, AS, and CP are all 128-bit wide and provide the buffering necessary to keep the instruction pipeline busy; the IB register is only 16 bits wide and is included solely to deal with data skew. Additional registers associated with the assembly include monitor registers that keep track of the valid instructions as they go through AS and CP, a counter that points to the next instruction parcel to be taken from CP, and various other registers (whose functions are discussed below) in the Unpack Record unit.

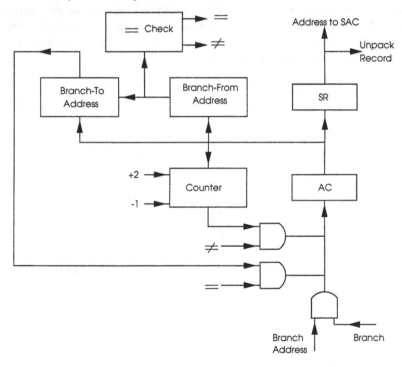

Figure 4.23.   MU5 branch target buffer

The MU5 BTAB has the straightforward design shown in Figure 4.23. It has an implicit single-bit history field and is used only for taken branches with an invariant branch-to field. As each new address from AC is sent to the Instruction Buffer Unit, it is also associatively compared with the addresses in the Branch-From-Address field of the BTAB. If there is a match, then the corresponding branch-to address is read out and sent to the SAC to fetch the instructions at the branch target, which instructions are then forwarded to the rest of the pipeline after the branch instruction. When the branch instruction enters the pipeline — the control point is several stages down in the pipeline — it is tagged with a bit indicating that the instructions following it may not be from the correct path; that is, that the prediction may have been incorrect. Subsequently, if the prediction turns out to have been incorrect, then the instructions after the branch are abandoned and a request made to the SAC for the correct sequence. At the same time, if the branch should have been taken but was not, then a line (selected using a Not-Recently-Used algorithm) of the BTAB is loaded with the branch-from and branch-to addresses. If a branch instruction has no entry in the BTAB, then the prediction is that it will not be taken: given the fact that the BTAB is accessed before decoding (and only a hit identifies an instruction as a branch) and the late location of the control point, no other type of default prediction is possible (unless a separate adder is used early in the pipeline).

Although the MU5 system was not designed to provide loop-catching capabilities, the fact that most of the machine's instructions are only one parcel wide means that there will be instances when both the branch-from and branch-to addresses are in the same 128-bit word. In such a case a Same-Word bit in the Unpack Record is set, and no requests are made to the SAC.

The BTAB here gives an example of how the architecture (instruction set) of a machine can cause complications in the implementation. In the case at hand the difficulty arises from having variable-length instruction formats: Consider the case where an incorrect prediction has been made and a new pair of values is to be entered in the BTAB. For a multi-parcel instruction, the program counter will point to the address of the first parcel, whereas the branch-from address should correspond to last parcel; the branch-from address is therefore not readily available. This situation is handled as follows. The value in the program counter value after a branch instruction is either *branch-from address + 1* or *branch-to address*. Since the branch-to address can be definitely known only after the branch condition has been evaluated, but the *branch-from address + 1* can be computed immediately, the control adder can produce both addresses without a loss of performance and then send them to the Instruction Fetch Unit. There, the first address is loaded into the AC register, decremented (hence the "-1" in Figure 4.23), and loaded into a line of the branch-from-address store; the second address is loaded (via the AC) into a line of the branch-to-address store and also sent (via SR) to the SAC, which then fetches the instructions at the branch-target.

A timing problem also occurred in the implementation of the BTAB: The comparison of addresses from the AC register with the addresses in the branch-from-address field normally proceeds concurrently with the operation of the counter in order to match the pipeline cycle time. But in the event of an equivalence, the correct address is that from the branch-to-address field, and this requires an additional cycle for the reading. To avoid a drop in performance, advantage is taken of the fact that two consecutive addresses from the AC differ only in the last bit, and by not presenting that bit during association, comparison can be simultaneously carried out for two addresses; the omitted bit is saved and later used to ascertain whether any equivalence is genuine. The size of the BTAB was determined by technological constraints (what could be implemented on a single logic platter) and by cost/performance trade-offs: doubling the size produced a very small improvement in performance.

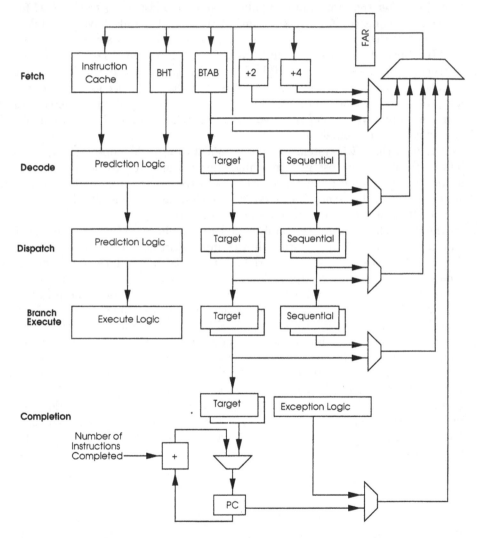

Figure 4.24.   IBM PowerPC 604 instruction-address-generation pipeline

4.4.3.4 PowerPC 604 and PowerPC 620 instruction-fetch/branch-prediction.   The PowerPC 604 is one of several high-performance implementations of the PowerPC architecture (see Section 1.5.5). In these implemen-

tations, the branching problem is tackled in a very aggressive manner that combines in a single implementation the use of a BTAB, a BHT, and conditional execution, with processing beyond several unresolved branches.

The PowerPC 604 BTAB consists of 64 lines and is fully associative. It has an implicit single-bit history field and is used only for unconditional and PC-relative taken branches (conditional and unconditional). Branches for which there are no entries are predicted "not-taken". The BHT consists of 512 lines of 2-bit saturating counters and is direct-mapped. For each branch, the BTAB is used to provide a (relatively) crude initial prediction of the direction, and the BHT is then used to obtain a more accurate prediction for unresolved conditional branches. The use of the BTAB to make an initial prediction is motivated by the fact that even in the best circumstances, prediction using only the BHT incurs a penalty of one cycle in fetching the target instruction; on the other hand, using the BTAB allows the fetching of the target to start earlier, thus eliminating the one-cycle delay for all but incorrect predictions. Since the BHT (because of its size and wider history field) is more accurate that the BTAB, in the event that there is a disagreement between the two predictions, that of the BHT takes precedence and is used to update the BTAB. (It may happen that the BTAB makes the correct prediction and the BHT makes the incorrect one, but this is rare; see Table 4.8.)

The instruction-address generation pipeline of the PowerPC 604 is shown in Figure 4.24. In the Fetch stage, the next block of four instructions to be processed is read from the cache. Instructions are decoded in the next stage and in the third stage are dispatched for execution. The Branch-Execute stage completes the processing of branch instructions and therefore ascertains the correct target address of a branch instruction. The function of the Completion stage include updating the program counter — to the address of a taken branch or by the number of completed instructions — and detecting exceptions.

In every cycle, each of the five main pipeline stages generates an address, one of which is selected, into the Fetch-Ahead Register (FAR), as the address of the next instruction-block to be processed; since the later a stage is, the older the corresponding instructions, the selection is made by giving priority in an increasing order from the last stage to the first stage. The address generated by the Fetch stage is either the result of incrementing the last fetch-address by the number (two or four) of instructions completed or skipped in the last instruction-block, or it is a branch-to address from the BTAB. The address from each of the Decode, Dispatch, and Branch-Execute stages is either the target address of a branch instruction or the next address in the straight-ahead path. And the address from the Completion stage is either a new value of the program counter, calculated as described above, or the address of an exception-handling routine. Each of the last four stages can handle up to two branch instructions in each cycle and, therefore, each has two pairs of registers (one per instruction) to hold both the branch address and the straight-ahead address.

The flow through pipeline is as follows. In the first stage, the address in the Fetch-Ahead Register is used to access the instruction cache (which returns a four-instruction block), the BTAB, and the BHT. The four instructions and, in the event of hits, the contents of two lines of the prediction tables are then forwarded to the Decode stage. The next step is the decoding of the four instructions and additional branch prediction (using the BHT value) in the Decode and Dispatch stages; this, more accurate, prediction may lead to a correction of the BTAB prediction, with a corresponding updating of the BTAB entry and a change in the predicted target address. Also, for unconditionals for which there are no BTAB entries, the Decode stage supplies a target address. Branch instructions are then executed, with either confirmation or corrections of earlier predictions. Lastly, the Completion stage alters the program counter for each completed instruction.

The branch prediction system implemented in the PowerPC 620 is similar to that of the PowerPC 604, except in the use of a return-address stack for subroutine jumps and in the size of the prediction tables: the BTAB has 256 lines and is two-way set-associative, and the BHT has 2048 lines (still direct-mapped). The results of a performance-evaluation of the PowerPC 620 system on the SPEC92 benchmarks are shown in Table 4.8.

Table 4.8.   Accuracy (%) of PowerPC 620 branch prediction

Branch		SPEC Program						
Branch		compr.	eqntott	espr.	li	alvinn	hydr.	tomc.
Branch	untaken	40.35	31.84	40.05	33.09	6.38	17.51	6.12
Result	taken	59.65	68.16	59.95	66.91	93.62	82.49	93.88
BTAB	correct	84.10	82.64	81.99	74.70	94.49	88.31	93.31
	incorrect	15.90	17.36	18.01	25.30	5.51	11.69	6.69
BHT	resolved*	19.71	18.30	17.09	28.83	17.49	26.18	45.39
	correct	68.86	72.16	72.27	62.45	81.58	68.00	52.56
	incorrect	11.43	9.54	10.64	8.72	0.92	5.82	2.05
Only BHT correct		0.01	0.79	1.13	7.78	0.07	0.19	0.00
Only BTAB correct		0.00	0.12	0.37	0.26	0.00	0.08	0.00
Overall Accuracy		88.57	90.46	89.36	91.28	99.07	94.18	97.95

*resolved=outcome known beforehand

4.4.3.5   SPARC64 fetch and branch units.   The SPARC64 is a recent 64-bit superscalar implementation of the SPARC architecture. The arrangement of the units that deliver instructions to the rest of the pipeline for execution is shown in Figure 4.25.

The Fetch Unit is the interface between off-chip cache and memory and the rest of the pipeline. The main function of this unit is to prefetch instructions (16 at a time, in two 64-byte lines), recode them, and store them in the on-chip cache and the Prefetch Buffers. For a branch instruction, the recoding forms an important part of speeding up the processing of branch instructions: the Recode Unit calculates a partial target address and stores this as part of the original instruction; after this preliminary calculation, all that is required to complete the calculation of the target address is a simple increment or decrement of the upper bits of the program counter.

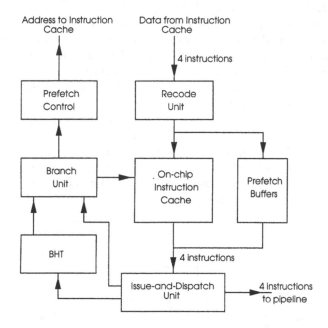

Figure 4.25.   SPARC64 instruction delivery

The Branch Unit completes target-address calculation for all branch instructions, and up to 16 unresolved branches can be in progress at a given moment. As instructions leave the on-chip cache or the Prefetch Buffer, they are decoded to detect branch instructions. If a branch instruction that is not a subroutine jump or return is detected (in a four-instruction block), then the lower 12 bits of the fetch-ahead counter (with the lowermost two bits masked out, leaving only ten bits) are used to index the BHT, which consists of 1024 2-bit saturating counters for the prediction history. Each line of the BHT corresponds to a line of the direct-mapped on-chip cache, and the four instructions in a line are fetched if the corresponding BHT counter produces a "taken" prediction. To facilitate quick recovery in the case of mispredictions, a separate register holds the address of the alternate path. For subroutine branches, a Return Prediction Table maintains a stack of return addresses that are used as predicted target addresses on subroutine returns.

4.4.3.6   DEC Alpha 21164 and 21264 branch prediction.   The DEC Alpha 21164 and 21264 are introduced in Section 1.5.4. The Branch logic in the Alpha 21164 instruction pre-processing pipeline (see Figure 4.5) examines instructions out of the second stage of the pipeline (i.e. from the cache or Refill Buffer), computes a new address for the next instructions to be fetched, and makes predictions of the targets of conditional branches.

The branch prediction table in the Alpha 21164 consists of 2048 lines of 2-bit saturating counters and is used in association with the instruction cache. (The use of two history bits, compared with just one bit in the Alpha 21064, was found to reduce, by half, the number of mispredictions in typical loop branches.) The table is addressed with the low-order bits of the program counter, as the instructions for which it is used are PC-relative branches; for full-address jumps, a mechanism similar to that used for variable return-target (discussed below) is used, since the architecture allows part of the cache target-address to be encoded in the instruction itself.

For branches with a variable target (i.e. subroutine returns) a 12-line Return-Address Stack is used. Full addresses are 43 bits, but each entry in the stack is only 11 bits, which is all that is needed to address the 8Kb direct-mapped on-chip instruction cache; the remaining 32 bits of the target address are taken from the tag field (i.e. the address field) in the line of the instruction cache addressed by the entry from the stack.

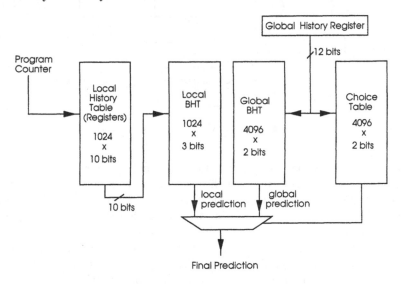

Figure 4.26.   DEC Alpha 21264 branch prediction

The branch prediction system in the Alpha 21264, the successor to the Alpha 21164, is one of the most sophisticated ones implemented to date (in any machine); it too is used in association with the instruction cache. The predictor is a hybrid one that combines local and global two-level prediction, in the organization shown in Figure 4.26. Three prediction tables are used: one for local prediction, based on the program-counter value and local history; one for global prediction, based on global history; and one that keeps track of the history differences between the local and global predictors. For each branch, two predictions (one local and one global) are made, and the third table is then used to select the better one of the two. And as in the Alpha 21264, a return-address stack is used to predict the targets of subroutines. The stack has a capacity for 32 entries and is located near the instruction cache so that it can be accessed in parallel and the output address quickly routed back into the cache.

The output of the branch prediction system is used by the Instruction Fetch Unit (Figure 1.11) to determine the addresses from which the next instructions in the pipeline are to be fetched. Each line of the instruction cache holds four instructions, a "next-line" address (for the subsequent instructions), and a "next-set" identifier. As the contents of a line are read out, the values in the last two fields are used to start the next cache access. The two fields initially point to the instructions at the next sequential address and are later updated according to the branch target-address when a branch is predicted as taken. Thus if the contents of a line that has just been read out include a branch instruction that is predicted taken, then the next-line and next-set fields identify the set and line containing the target instruction; otherwise, they point to the next instructions on the straight-ahead path. The estimated average accuracy of the system is 95% on SPEC95 benchmarks.

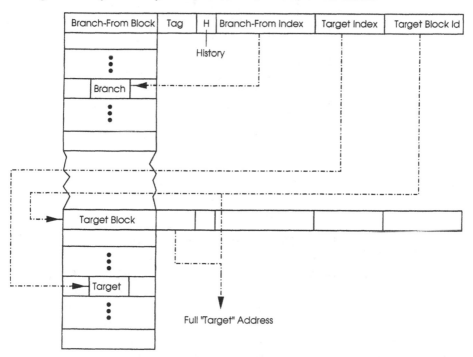

Figure 4.27.   AMD K5 branch prediction

4.4.3.7 AMD K5 and AMD K6.

The AMD K5 branch prediction system combines the essential elements of both the BTAB and the BTIB in a way that gives high performance at a reasonable cost: The branch logic stores only the addresses of the branch and its target. But since the high-order bits of the instruction cache tags match the high-order bits of the full branch-from and branch-to addresses, the actual branch-from address stored is just an index into a block in the cache, and the stored branch-to address consists of just a pointer to the target block and an index into that block. Evidently the stored addresses are much smaller than full memory addresses. This arrangement is shown in Figure 4.27. The prediction itself is dynamic, with a 1-bit history field.

There are two cases that need special consideration in such an implementation: one is that the actual target address may change, and the other is that the target block may have been replaced. Both cases are resolved by computing a full address from the tag in the indexed target block and sending this along with the branch instruction for confirmation of correctness when the branch is executed.

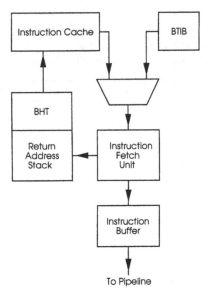

Figure 4.28.   AMD K6 instruction fetch

The AMD K6 is introduced in Section 1.5.7. It is an example of a state-of-the-art microprocessor that employs a range of sophisticated branch handling techniques: the branch logic consists of a BHT, a BTIB, a Return-Address Stack, and a dedicated Branch-Execution unit. The organization of the branch logic is shown in Figure 4.28. The BHT has 8192 entries and is used in the implementation of a two-level adaptive predictor: the contents of a 9-bit branch-history register are concatenated with 4 bits of the program counter to form a 13-bit index into the table. (The addresses required to fetch the target-instructions of branches predicted "taken" are computed by special adders in the Decode stage.) The BTIB has 16 entries, each of which stores the first sixteen bytes of instructions at the target address. And the Return-Address Stack has 16 entries for addresses and functions as a cache for return addresses that have been pushed into a stack in memory.

Conditional branches are resolved in the Branch Execution unit, which is designed to support speculative execution of instructions following a branch, to such an extent that there can be up to seven unresolved branch instructions in progress at a given moment. The processing of unconditional branches does not incur any penalty, as they are decoded early and the target addresses calculated early enough.

4.4.4 Summary

In this section we have discussed the use of branch prediction as a solution to the problem of branches. The emphasis has been on hardware branch prediction, the technique most used in current very-high-performance pipelined machines. Various aspects of the design of hardware branch prediction units have been discussed. Unlike the case for an ordinary cache, hit rate alone is not a good performance indicator in branch prediction: what matters, at the very least, are both the hit rate *and* the proportion of correct predictions among these hits. Beyond these other important factors include misprediction penalty, instruction-set architecture, languages and compilers, default prediction for misses, particular pipeline implementation, and so forth. As an example of why it may be insufficient to consider just one aspect of performance, we cite the case of the Manchester MU5: The very small BTB has an average prediction accuracy that is about 70% at best and much lower in some cases, depending on the language and compiler used and the nature of the task (compilation or execution). This might be considered rather low, but when account is taken of the fact that the number of times in which a branch instruction is followed by the correct sequence of instructions when the BTB is used is three to four times higher than when it is not, then the effectiveness of the BTB becomes evident. In the latest machines, performance figures such as those in Table 4.8 are typical.

Hardware branch prediction was first implemented in the Manchester MU5, but for nearly a decade after the commissioning of the that machine no other machine implemented hardware branch prediction. This has now changed, and hardware branch prediction is implemented in a number high-performance machines, including vector supercomputers, such as the IBM 3090 and the NEC SX machines. Many of the current high-performance machines take an aggressive approach to dealing with the branch problem and use branch prediction in combination with other techniques described above. The sophistication of implementations has continued to grow, and although improvements in implementation may have only small effects on prediction accuracy, they are still worthwhile if the branch costs are large. It is therefore to be expected that further improvements will take place and that such techniques will spread to machines at the low end of the performance spectrum.

Table 4.9.   Instruction predication

	Original Code	Predicated Code
	MOV R1, 0	MOV R1, 0
	MOV R2, 0	MOV R2, 0
	LD R3, ADDRA	LD R3, ADDRA
L1:	LD R4, MEM(R3+R2)	LD, R4, MEM(R3+R2)
	BGT R4, 50, L2	PRED-GT P1, P2, R4, 50
	ADD R5, R5, 1	ADD R5, R5, (P1)
	JMP L3	ADD R6, R6, (P2)
L2:	ADD R6, R6, R1	ADD R1, R1, 1
L3:	ADD R1, R1, 1	ADD R2, R2, 4
	BLT R1, 100, L1	BLT R1, 100, L1

4.5   PREDICATED EXECUTION

In *predication* the view is taken that the way to deal with branches is to eliminate them from the instruction stream, and the underlying architecture is then designed accordingly [73, 74, 117]. An instruction is associated with a condition (predicate) whose value determines whether or not the processing of the instruction eventually has effect on the architectural machine-state. Conditional branches can then be eliminated by predicating instructions according to which outcome of the condition with which they are associated: one predicate is associated with each of the two paths from a branch-fork. As an example, Table 4.9 shows the predicated and non-predicated assembly-language code produced for the code fragment

$$\textbf{for } \text{i:=1 to 99 do}$$
$$\quad \textbf{if } A[i] \leq 50 \textbf{ do}$$
$$\quad\quad \text{j:=j+1}$$
$$\quad \textbf{else do}$$
$$\quad\quad \text{k:=k+1}$$

Here the PRED-GT instruction sets to 1 the 1-bit predicate register P1 if the value in r4 is greater than 50 and to 0 otherwise and also sets the 1-bit predicate register P2 to the complement of the value in P1. The following instructions are then executed, but the predicated ones are not allowed to alter the named destination register unless the corresponding predicate is 1; effectively, the execution of an instruction that is predicated on 0 converts it into a NOP. Thus the effect of predication in a piece of code with multiple control paths is to permit, by exposure of more execution threads, the exploitation of more instruction-level parallelism that would otherwise be the case with just one thread.

Table 4.10.   Instruction predication in HP-PA

HP-PA Code	Comment
SUB, $\geq$, R1, R2, R0	Subtract contents of R2 from contents of R1, result to R0. If value in R0 is ≥ 0, then nullify next instruction.
SUB R2, R1, R0	

Predication may be *restricted* or *unrestricted*, according to whether or not only a few or all of the instructions in the architecture are predicated. Thus, for example, Cydrome's Cydra 5 supports has unrestricted predication, in which each instruction names one of 128 1-bit predicate registers [96]; the HP Precision Architecture has restricted predication in which only branches and data transformation instructions are predicated [70]; and the the DEC Alpha has even more restricted predication — only on a MOVE instruction [101]. Unrestricted predication may be implemented by either explicitly including an additional operand (the predicate bit) in each instruction or by using special instructions to control the execution of (following) instructions that are not explicitly predicated.

The restricted predication of the HP Precision Architecture is essentially a feature that is intended to support the one-cycle delayed branching of the architecture: it allows any branch or data-transformation instruction to squash ("nullify") the execution of the instruction immediately after it. As an example, the compiled predicated code to compute into register R0 the absolute difference of the values in registers R1 and R2 is shown in Table 4.10: the predication has eliminated the branch that would otherwise be required for the implicit IF-ELSE construct. For branching, the predication is used as follows. The immediately following instruction (i.e. the one in the delay slot) is nullified for backward branches only if the condition is false and for forward branches only if the condition is true. Since the architecture uses the static branch prediction backward–taken/forward-not-taken, the first instruction in a loop body can be moved out, thus ensuring that the delay slot is always filled. Thus, for example, given a code fragment with the logical structure

> **repeat**
> > *instruction-1*
> > *instruction-2*
> > $\cdots$
> > *instruction-k*
> **until** C
> > *instruction-m*

the actual compiled code on the HP would have the form

instruction-1
L1: *instruction-2*
   ...

instruction-k
COMBT, C, N, R1, R2, L1
instruction-1
instruction-m

The loop-closing branch, COMBT is predicated so that if condition C is true, then *instruction-1* after the COMBT is executed and control transfers back to to L1; otherwise *instruction-1* is nullified and processing continues with *instruction-m*. For a forward branch, the predication shortens the code by one instruction in IF-ELSE constructs.

Predication, even unrestricted, by itself does not completely solve the branching problem, nor is it without cost: First, not all branch instructions can be eliminated by predication, loop-closing branches being an example. Second, instructions that are nullified on predication may not change state, but they are processed and executed in the same way as other instructions that do useful work; from a practical point of view, keeping the pipeline busy with instructions that do nothing is not much better than keeping it empty. The extra work may be likened to the misprediction penalty when just branch prediction is used; however, given the total cost of mispredictions in deep or wide pipelines, predication is likely to be less costly. (Note that predication may also be used in conjunction with prediction.) Nevertheless predication has been shown to be useful in the removal of difficult-to-predict branches and in increasing average distances between branches (and, therefore, between mispredicted branches), and the challenge is to (with compiler assistance) ensure that the gains from increased exposure of instruction-level parallelism outweigh the loss from any wasted efforts. One significant use of predication that is likely to influence future machine designs is in the forthcoming implementations (Merced and Mckinley) of Intel's IA-64 architecture.

4.6   OTHER SOLUTIONS OF THE BRANCHING PROBLEM

There have been a number of other proposals for solutions of the branching problem. Most of these have as yet not found widespread application, and in this section, we shall briefly describe only two of the more promising ones: processing along both paths from the branch and processing instructions from several independent streams.

Dual-path processing

In principle, one could deal with the control-transfer problem by concurrently processing instructions from both the straight-ahead path and the target path and then discarding one path when the branch is finally resolved. But practical implementation of this scheme presents several difficulties: First, in order to avoid too much duplication of hardware, it requires that the target address always be known early, which may be difficult with a late control point. Second, in order to simultaneously support instruction-fetch for both control paths, in addition to data accesses by the processor, it requires that the memory unit have greater performance than would otherwise be the necessary. Third, the possibilities of further branching within the initial two paths means that, in general, facilities must be provided to process a number of paths, but this number grows exponentially with the degree of forking. Fourth, even if the technique is limited to just two paths, it may require the replication and coordination of an impractically large amount of hardware.

Nevertheless, a very limited variant of the scheme has been implemented: We have seen an example above, in the Motorola 68040. The technique is also used in some IBM machines, such as the IBM 370/168 and the IBM 3033, in which the (small) instruction buffers are replicated, and on a branch, instructions in both directions are fetched into the buffers; those instructions on the straight-ahead path are conditionally processed until the branch is resolved and then thrown away (and replaced with those in the alternate buffer) if the branch should have taken place. And in the Motorola DSP96002, instructions along both paths are fetched, but only those on the taken path are decoded and processed. Similar systems are used in the IBM POWER1 and POWER2 (two implementations of the POWER architecture). Another example of the use of this technique is in the Cyrix 6x86, which is interesting because it also incorporates a BTB: instructions are fetched according to a prediction, but for fast recovery in the event of a misprediction, some instructions are also fetched along the alternate path [77]; a similar system is implemented in the NEC V80 [64].

Multithreading

Suppose we have an n-stage pipeline and at least n active instruction-streams (i.e. instruction *threads*).[9] If we repeatedly cycle through these threads, and in each pipeline cycle introduce into the pipeline an instruction from a different thread, then at any given moment all the instructions in the pipeline will be independent. This is known as *multithreading*, and it solves the branching problem by ensuring that the pipeline is kept busy processing instructions from other threads while one thread is held up on a control-transfer.[10] In a straightforward implementation, multithreading may be realized by a complete replication of appropriate hardware, which can be costly. This, however, need not be the case, since partial multithreading can still be effectively used in conjunction with the other techniques described above, resulting in a more efficient implementation. The design of multithreaded computers is currently a very active area in computer-architecture research, and we may expect the technique

---

[9]This is simplistic, since there are some operations have multi-cyle latencies.

[10]Multithreading also solves other problems that can reduce the performance of a pipeline: for example, that of data hazards between instructions in a pipelines.

to be used in more machines in the future. A good introduction to the field will be found in [53].

4.7   EFFECT OF INSTRUCTION-SET ARCHITECTURE

We shall now briefly comment on some of the effects that a machine's architecture can have on the branching problem. The aspects of the instruction set that are discussed are the instruction format, the encoding of branch addresses, the choice of control-transfer instructions, and the use of condition codes.

Instruction format

Ideally, the instruction set should be based on a few simple formats of fixed length. Complex formats can affect the implementation in at least two ways: The first is that they make it hard to detect branch instructions early in the processing, and furthermore, alignment and decoding times are added to the branch latency. The second is that they increase the cost and complexity of the hardware. An example of the latter in the MU5 BTB implementation (described above); another example is in the VAX 9000 BTB, in which each line of the BTB has six bits to indicate instruction-length, with the bits being used in a relatively complex manner.

Encoding of target address

We have already seen above that some types of target-address encoding can help with branching, by shortening the time required to calculate the target address. Indirect addressing, for example, is undesirable in so far as determining the target normally requires an operand to be read at what can be a relatively late stage in the pipeline. There have been a variety of suggestions for encodings that facilitate fast target-address calculation. Typically, the goal in these encodings is to eliminate control addition or to, at least, limit it to just part of the address. One example of such encoding is in the DEC Alpha AXP architecture, in which use is made of otherwise free bits in the instruction format to partially specify the target-address for a full-address jump; this partial address is sufficient to start the accessing of the fastest (lowest) of level of the instruction cache, thus speeding up the process of getting the target instructions into the pipeline. Fast address calculation is useful even in machines that use hardware branch prediction: if, say, a BTB does not have an entry for a branch that is taken, or, if the prediction is incorrect, it helps to be able to quickly calculate the target address.

Choice of instructions

Certain types of both branch and non-branch instructions can help in ensuring that instructions are fed into the pipeline at an adequate rate. For example, the use of conditional-move instructions (with a generalization to full predication) can reduce the number of branch instructions used in a program and software can also help here in reducing the number of branch instructions if the compiler is used to unroll loops. Another instance where the choice of instructions is crucial is in the addressing modes used: complicated addressing increases the time required to calculate target address and also affects the position of the control point and branch prediction mechanisms.

Condition codes

In a typical architecture, condition codes are normally set, both explicitly and as a side-effect, by any number of instructions; this can be problematic for a pipelined machine and in other respects as well [97].

A common way to realize conditional branching in an architecture with condition codes is with two instructions: one that sets a condition code and another that tests the condition code and transfers control. Although explicit condition codes are helpful, in that they facilitate branch spreading, whence their retention in some modern architectures, the ability to arbitrarily set them (especially as a side effect) is not: since any instruction that is used to separate the two instructions must not be one that can alter the condition code to be tested, the number of instructions that can be used to achieve the spreading is reduced, even though there may be several usable instructions that are not related to the branching. Even if branch spreading is not used, a problem still exists — that instructions following the branch instruction in the pipeline may change the condition codes before the testing, even though the branching might not otherwise be dependent on such instructions. Moreover, these difficulties are aggravated if condition codes can be set as a side-effect or if instructions are processed out-of-order. One solution to such problems is to stall the pipeline whenever the processing of a conditional branch instruction is started and there exists an uncompleted instruction that can change the condition codes; that is, it is assumed that *any* such instruction can affect the branching. But this is undesirable from a performance viewpoint, and, consequently, there is likely to be an increase in the complexity of the hardware if both performance and correctness are to be maintained. Examples of such complications will be found in the IBM RS/6000, the PowerPC 601, and the PowerPC 604: in the PowerPC 601, for example, copies of the Condition Register are required at seven points in the pipeline, and their coordination and use is rather complex; and in the PowerPC 604, the use of condition codes requires the serialization of some instructions that could otherwise be executed in parallel with others. Another drawback of condition codes is that they form part of the machine-state that must be saved and restored on interrupts.

Because of the difficulties above, some architectures avoid the use of condition codes and instead realise branching with single instructions that combine the testing and transfer of control; others have separate branch instructions that test only the contents of general-purpose registers, i.e. conditions cannot be set as a side-effect. The former has the additional advantage of reducing both the static and dynamic instruction counts; the latter allows branch spreading. The MIPS, DEC Alpha AXP, HP Precision, and Motorola 88000 are examples of architectures in which condition codes have been eliminated for these reasons

[2, 27, 69, 80, 101]. If condition codes are retained in an architecture, then the problems mentioned can be lessened by requiring that they be set only explicitly and by only a few types of instructions. It should also be noted that in the implementation of an architecture with condition codes, the actual delay involved in reading the condition-code register can be eliminated by adding logic to directly forward comparison results to the condition-evaluator at the same time as they are written into the register; such a system is used in the PowerPC 601 [7].

Self-modifying code

An architecture that permits self-modifying code presents a problem, with respect to branching, since it can be the case that the target instructions of a branch are being fetched while WRITES modifying some of them are still underway, thus creating what might be considered a control-data hazard. The checks and coordination required to ensure correctness in such a case can affect the complexity and performance of the branch-microarchitecture. Nevertheless, some machines, such as the Intel Pentium, permit self-modifying code and have in place mechanisms to ensure proper processing [1].

4.8 SUMMARY

We have discussed a number of ways to deal with the difficulties caused by control-transfer instructions in trying to ensure that an instruction pipeline is supplied with instructions at a sufficiently high rate. The methods discussed range from purely software-implemented (with architectural support) ones that on their own are used in the simplest machines, through simple hardware instruction buffers, to sophisticated hardware-implemented branch-target prediction techniques. A number of other related issues have also been discussed.

As technology continues to improve, and machines at the lower ends of the performance range adopt the techniques currently used in machines of the highest performance, we may expect improvements in these techniques, as well their widespread use. Currently, low-performance machines use only one or two of the simpler techniques discussed, while the high-performance machines implement a combination of all or almost all of them.

5 DATA FLOW: DETECTING AND RESOLVING DATA HAZARDS

A pipeline can fail to achieve its maximum speedup if there are discontinuities in the supply of instructions or data. Discontinuities in the flow of instructions have been covered in the preceding chapter; in this chapter, we shall discuss the problem of discontinuities in the flow of data as well as corresponding solutions. The data-flow discontinuities arise mainly from two sources: one is a mismatch between the rate at which the pipeline requests data and the rate at which the data is delivered to the pipeline; the other is *data hazards* (or *data dependences*) that occur between instructions in the pipeline when one instruction cannot proceed because its progress depends on that of another instruction. The basic problem of mismatching rates is largely solved by the use of appropriate high-speed intermediate storage (cache, registers, etc.) and other techniques discussed in Chapter 3 and will not be considered further. This chapter is therefore devoted to just the hazards and related issues.

In the first section of the chapter, we describe the three types of data hazards that can occur in a pipeline and briefly discuss how to resolve them; two of the three types of hazard may be viewed as artificial and can be eliminated by appropriate use of hardware or software, whereas dealing with the third is more difficult. In the second section, we discuss the technique of *renaming*, which is useful for eliminating the artificial hazards. The third section deals with techniques used to speed up the execution of instructions that are involved in genuine (i.e. non-artificial) hazards — essentially, these techniques involve organizing the processing in such a way that instructions that are unable to progress do not necessarily hold up the execution of unrelated instructions, which can then be executed as soon as possible. The fourth section consists of a number of case studies, and the fifth is a summary.

5.1   TYPES OF DATA HAZARDS

In this section we shall identify and describe the three types of data hazards that can occur in a pipeline and the conditions under which they occur. We shall also briefly describe techniques to detect and resolve them; later parts of the chapter will deal in detail with how to implement these techniques.

The three main types of data hazards are exemplified by the code fragments in Table 5.1; in all three examples, the hazard is on R0. The first type of hazard is *read-after-write* (RAW) (or *input-output dependence*), in which an instruction needs to read from a storage location (memory, cache line, register, etc.) that is yet to be written into by an uncompleted instruction that appears earlier in the instruction stream. The second type of hazard is *write-after-write* (WAW) (or *output dependence*), in which an instruction needs to write to a storage location that is also the target for writing by an uncompleted instruction that appears earlier in the instruction stream. (In the example of Table 5.1, it is assumed that division takes much longer than addition.) And the third type of hazard is *write-after-read* (WAR) *hazard* (or *anti-dependence*), in which an instruction needs to write to a storage location that is yet to be read from an instruction that appears earlier in the instruction stream. Each of these three cases implies that proper sequencing be maintained if the intended results are to be obtained.

Table 5.1.   Types of data hazard

Read-After-Write	Write-After-Write	Write-After-Read
R0 ⇐ R1*R2	R0 ⇐ R1/R2	R1 ⇐ R0*R2
R4 ⇐ R0+R3	R0 ⇐ R3+R4	R0 ⇐ R3+R4
(a)	(b)	(c)

Although we shall be mainly concerned with the occurrence of these hazards in the context of computational instructions, it should be noted that other sequencing problems that occur elsewhere in a pipeline may also be viewed in terms of hazards: for example, the problem of condition codes in branching is essentially a RAW hazard on a condition code, and the ability of multiple instructions in progress to set a condition code is essentially a WAW hazard on the condition code. (These hazards in the flow of control are also sometimes referred to as *control hazards*.) Additional remarks are also in order on the WAW hazards: Although the second example in Table 5.1 may seem contrived or trivial, three things should be noted about it: first, such a hazard can occur because there are two or more WRITE stages in a pipeline; second, a non-

trivial (*second-order*) WAW hazard can occur because there is an intervening
READ, between the two WRITEs, that gives rise to a RAW and WAR hazards
— for example, if the instruction R6 $\Leftarrow$ R0*R5 were to appear between the two
instructions in the example of Table 5.1(b); third, as odd or improbable as the
first-order hazard (i.e. of the type given in the table) may seem, the hardware
designer must nevertheless ensure that the sequencing indicated by the program
is carried out. Lastly, it should be noted that a WAW hazard can occur even
with just one instruction: for example, if an instruction occurs in a loop and
one instance of it is initiated before an earlier instance is still underway, then
a hazard can arise between the two instances. The other two-types of hazard
can similarly be *loop-carried*

As the examples in Table 5.1 show, the order in which instructions are issued
and executed largely determines the degree to which hazards exist and how they
can be resolved: if the instructions are issued and executed in an order that is
in strict keeping with their initiation, then hazards are easily resolved simply
by stalling the second of the two instructions involved in a hazard until the
first has been executed: for example, given the code fragment in Table 5.1(c)
case in Table 5.1, with in-order processing the contents of R0 will be have been
read before starting the second instruction, and therefore no problem will arise.
Thus, for example, in the MU5 primary pipeline (Section 5.4.1) and in the Intel
Pentium a WAR hazard is easily dealt with because READS are always processed
before WRITES. In superscalar machines, however, the main means for obtaining
high performance are the multiple execution (functional) units, and making the
best use of these implies that such machines must necessarily have out-of-order
processing of instructions. But with out-of-order processing of the code in Table
5.1(c), the addition might be completed before the multiplication is started, and
proper sequencing on the use of R0 is then required.

The main issues in dealing with hazards are how and when to detect them
and how to resolve them once they have been detected. The solutions may be
in hardware or software, but, in general, software solutions tend to be only partial. If the operands involved have explicit and invariant names — for example,
register names — then RAW and WAR hazards can be detected by (simultaneously) comparing (at compile-time or, in the hardware, at run-time) the
names of source operands in one instruction with those of destination operands
in preceding instructions, and WAW hazards can be detected by comparing
destination-operand names. (Note, however, that compile-time detection may
not always work on loop-carried hazards.) If, on the other hand, the operand
are implicit or have variable names — for example, if they are computed addresses — then software solutions are generally not applicable, and hardware
solutions may also be of limited usefulness.

As a simple example of hazard-detection by the comparison of register names,
Figure 5.1 shows the hardware structure used to detect register hazards in the
Advanced Micro Devices Am29000 microprocessor, a four-stage pipeline with
1-cycle execution [3]. The register file is accessed with an address that is either
the absolute register-address or the register-address relative to the stack

pointer. The logic shown detects three kinds of hazards: when a source-operand address of the current instruction matches the destination-operand address of the immediately preceding instruction, when a source-operand address matches the target address of an uncompleted LOAD, and when the destination address of the current instruction matches the target address of an uncompleted LOAD. (The register ETR holds the address of an outstanding LOAD until that operation is complete.) In the first case the data awaited is sent directly to the execution unit, instead of being read from the register file; in the second case, the data is sent directly to the execution unit if the data becomes available in the next cycle, otherwise the pipelines stalls until the LOAD is completed; and in the third case, the LOAD is prohibited from modifying the contents of the register file, as the operation will have been superseded by the completion of the current instruction. Similar, but more sophisticated, systems are implemented in many high-performance machines, and we shall examine these below.

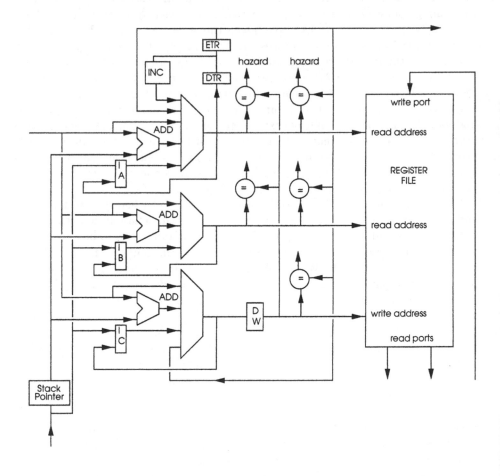

Figure 5.1.   Am29000 register address generator

The use of relatively complex logic (e.g. involving the use of comparators, as above) is not always necessary, depending on the pipeline design. For example, in a simple pipeline (e.g. one that stalls on hazards) two extra bits are sufficient for the detection of hazards: One bit is set to reserve a each register for writing and reset when the writing has taken place, and one bit is similarly used for reading. Hazards are then detected as follows. A RAW hazard is detected when an instruction needs the contents of a register that has been reserved for writing, a WAW hazard is detected when an instruction needs to write into a register that has been reserved for writing, and a WAR hazard is detected when an instruction needs to write into a register that has been reserved for reading. Mechanisms of this type have been implemented to varying degrees in a number of machines, such as the CDC 7600, the Manchester MU5 (Section 5.4.1), the CDC 6600 (Section 5.4.2), and the Motorola 88000 [5]. Evidently, the basic arrangement does not easily lend itself to implementations in which instructions are processed out-of-order or in which several instructions are issued in each cycle and in which, therefore, there can be dependences between a group of instructions under consideration for issue; Nevertheless elements of the basic idea will be found in a number of such machines.

As indicated above, software-implemented techniques may also be used to deal with hazards. An example of a simple compile-time (partial) solution to the problem of data-flow discontinuity is the *delayed loading* that has been used in some RISC processors. This is the data-flow analog of the *delayed branching* solution (see Chapter 4) to the control-flow problem in these processors. The basic idea is this: Suppose an instruction has as a source operand data that is not yet to be loaded from memory. If the delay (and pipeline stalling) implied is of a predictable duration, then it be eliminated by the placement of other independent (and, ideally, useful) instructions between the instruction loading the data and the instruction using it. Essentially, what we have here is a RAW hazard on a memory location, and similar techniques could be applied to other types of RAW hazards (e.g. on computational registers): by placing useful instructions between the READ and WRITE instructions in such a hazard, we achieve an effect similar to that of branch spreading (Chapter 4) and eliminate or reduce the seemingly inevitable waiting times. In general, *static instruction scheduling* is a body of compiler techniques that deal with hazards by re-arranging and modifying code obtained by straightforward means. Although such techniques have been quite successful with simple machines, their effectiveness is marginal with long pipelines, multiple functional units, branch prediction, and other sophisticated hardware implementations. Accordingly, the more sophisticated techniques discussed below all involve the run-time *dynamic instruction scheduling*, by the hardware, of instructions; of course, compile-time scheduling may be, and frequently is, used to supplement such techniques.

There are three main ways in which to handle hazards in hardware: the first is to stop the pipeline until no inconsistency can occur; the second is to allow instructions to proceed as far as possible without destroying consistency —

in the terminology of Chapter 1, to have out-of-order completion but ensure in-order retirement; and the third is to first eliminate as many hazards as possible and then revert to either or both of the first two approaches. The first approach clearly should be a measure of last resort, given that the objective in pipelining is high-performance, through a continual flow of instructions. The second technique is useful mainly with WAR and WAW hazards: For example, with the code fragment in Table 5.1(b), both the multiplication and addition could be carried out concurrently, but with the second instruction then not allowed to write into R0 until the first had done so; alternatively, results may be written in any order into temporary registers (used as working space by the execution units) and then later transferred (in correct order) into the named destination registers. And with the code fragment in Table 5.1(c), the addition could be carried out immediately, provided the writing into R0 is held up until the first instruction has read from it. The third technique relies on using new names for operands and, accordingly, is known as *renaming*. The essential idea in renaming is that by changing each destination-operand name and each subsequent reference to the same operand as a source, up to (but excluding) the next reference of the same name as a destination, all the artificial hazards are completely eliminated. As an example, the application of renaming the third and fourth references to R0 in the code fragment of Table 5.2(a), in which all three types of hazard occur, yields the code fragment in Table 5.2(b), in which only RAW hazards remain. Evidently, renaming can be done in hardware or software; however, renaming in hardware offers more flexibility, as it is effective on computed addresses as well, and, moreover, the names used need not be limited to those available to the programmer or compiler (i.e. to just the architectural names) but can be of any number that yields the best performance in the hardware.

Once renaming has been carried out on a group of instructions, only RAW hazards remain, and if instructions are to be processed in parallel or out-of-order, then it is necessary to have some indication of where the hazards occur. The most common method, as will be evident in the case studies below, is to associate a unique tag with (the writer of) a result and the same tag with all readers of the result; the tag is essentially a ticket that is used to locate, when a result is produced, all instructions that are awaiting the result. A different, and less commonly implemented, method is to use an *ordering matrix*: for an n-instruction group, the matrix is an n-bit$\times n$-bit table in which bit (i, j) is set if instruction i depends on the result of instruction j. An example of the use of an ordering matrix will be found in the memory-access system of the MIPS R10000.

In summary, we have in this section described the three types of hazards (read-after-write, write-after-read, and write-after-write) that can occur in a pipeline. We have also briefly indicated how they can be detected and resolved. In what follows, we shall discuss in detail the implementation of such detection and resolution mechanisms, as well as the elimination, where possible, of some hazards and how to speed up execution for those hazards that cannot be eliminated.

Table 5.2.   Renaming to eliminate hazards

Original code	Renamed code
R0 ⇐ R1/R2	R0 ⇐ R1/R2
R3 ⇐ R0*R4	R3 ⇐ R0*R4
R0 ⇐ R2+R5	R7 ⇐ R2+R5
R2 ⇐ R0−R3	R8 ⇐ R7−R3
(a)	(b)

5.2 IMPLEMENTING RENAMING

In this section, we shall first describe a basic implementation of renaming with registers and also show how a similar implementation can be used with a cache and memory addresses. We shall then describe an "optimized" register-renaming system that has similarities with the cache system.

In the example of Table 5.2(b), any free architectural register could have been used in place of R7. Furthermore, if the renaming is done in hardware, then the physical registers associated with the new names need not be visible to the programmer and can therefore be of any number. Accordingly, a typical implementation of renaming is based on viewing programmer-visible names as names of entries to a mapping table that holds pointers to the actual locations used to store data; thus there is a clear distinction made between the architectural registers and the physical registers. At any instant, an entry in the mapping table is free or points to an entry in the physical storage that holds (or will eventually hold) the data and has some indicator of which of the two is the case; each line of the data-storage table also has some indicator of its "free/empty" status, according to whether or not it contains valid data. The tables are then used as follows. As each instruction is processed, for each register that appears as the name of a destination operand, a free line in the data table is selected and the corresponding entry in the mapping table is set to point to that line. All subsequent instructions, up to the next one that use the same register as a destination operand, first access the mapping table and then replace the program-given name with the contents of the line in the mapping table, which address is then used to access the data table. Once all the READS corresponding to a particular WRITE have been carried out — for example, for R0, after the second instruction in Table 5.2(b) — the data-table line in question is declared free and available for use in another renaming operation. (This status can be detected by keeping track of the number of pending READS for each entry in the mapping table: a counter is incremented once for each pending READ and decremented once for each completed READ. An alternative to the use of counters is to arrange for data to be sent directly to waiting

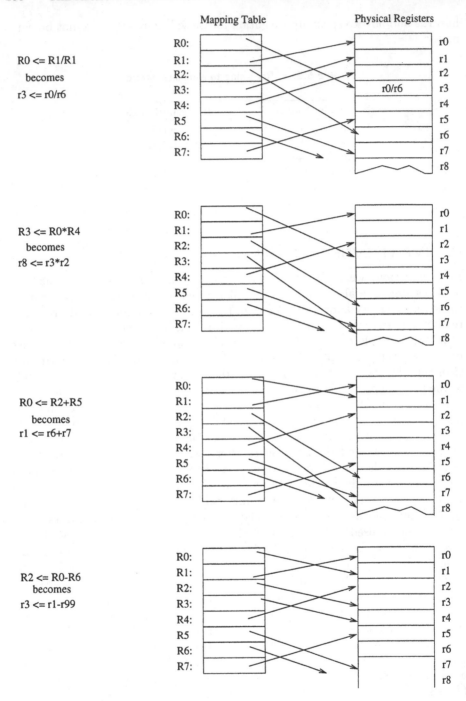

Figure 5.2.   Implementation of renaming

instructions, thus eliminating the need to read from the register file; this is discussed in more detail below. And a third alternative is the implementation in the CDC 6600, which is discussed below.) Figure 5.2 demonstrates this procedure for the program fragment of Table 5.2, assuming there are at seven program-visible registers (register names).

RAW hazards are easy to detect with the system just described: they occur when the status bit of a selected line in the data-storage table indicates that there is not yet any data in the line. In such a case, the address of that line (or some other unique identifier) is appended to the reader-instruction, which is then stalled until the data becomes available; the mechanism by which data and waiting instruction are finally brought together is described in Section 5.3. Implementations, with minor variants, of the type of renaming system just described will be found in in the IBM RS/6000 (Section 5.4.4), in the MIPS R10000 (Section 5.4.5), and in the DEC Alpha 21264.

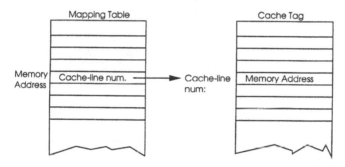

Figure 5.3.   Renaming with a cache

Hardware renaming, as described above, is easily extendible to a cache, memory buffer system, and other similar units in which the names can be computed addresses, providing a suitable ordering is maintained, where necessary, in processing WRITEs to memory. Consider, for example, a fully associative cache. If we added a mapping table as above, we would end up the situation shown in Figure 5.3, in which there is an obvious duplication of information: for each line, the address that labels the mapping table is the same address that appears in the corresponding cache line, and the address that labels the cache line is the same one that appears in mapping table entry for the address contained within the line. We may therefore eliminate the mapping table by using a bit to indicate that "the address contained within this line is currently mapped to this line"; that is, the address field of the cache functions as the mapping table. This "mapping"(M) bit is included in the usual associative comparison. An example is shown in Figure 5.4. (An alternative to using the mapping bit is to have age-priority logic that always identifies the most recent of several entries with the same tag-address.) The cache then processes each destination name (memory address) as follows. First, associative search is carried out. If non-equivalence occurs, then the address is entered into a newly selected cache line, and the corresponding mapping bit is set. If equivalence occurs, then the same process takes place, except that, in addition, the mapping bit for the old line on which equivalence occurred is set to 0. The original memory address in the instruction is then replaced with the address of the selected cache line (or with the data from the line). It will be seen that READs are handled correctly, since only last of any WRITEs will have the corresponding mapping (M) bit set, and, therefore, equivalence will occur on only that line. In this system, a RAW hazard is detected when a line with a matching address and a set mapping bit has a Data-Valid bit that indicates the absence of data; the situation is handled in the same way as in the register-renaming case. A cache-renaming system is described in more detail in [21]. In what follows, we shall use the term *renaming table*[1] to refer to the mapping-table/data-table combination, for both the explicit and implicit mapping table, with either registers or cache.

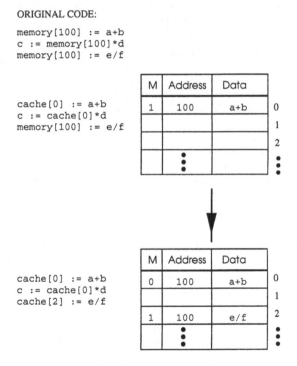

Figure 5.4.   Optimized renaming with a cache

The cache renaming system of Figure 5.4 differs from the register renaming system of Figure 5.2 in two essential ways: the first is that in the former the mapping table does not necessarily have the same number of slots as the data-storage table and is also much smaller than the corresponding name-space (i.e. the memory-address space); the second is that in the latter system *every* destination is renamed, which is not the case with the cache system. This

---

[1] A register renaming table is also sometimes referred to as a *register alias file*.

suggests that the register-renaming system can be optimized in two ways: first, the data-storage table need only be as large as the maximum possible number of outstanding WRITEs; second, the renaming need only be done for destination names that are currently is use. The maximum possible number of outstanding WRITEs depends on the longest pipeline path beyond the renaming stage and, therefore, on the particular implementation. Consequently, for a given machine, the smallest possible data table can have fewer or more slots than the number of architectural registers; for example in the PowerPC 604, a total of 17 renaming slots are used for a total of 65 architectural registers. If the data-storage table has fewer slots than there are architectural registers, then the mapping table also need not be any larger; furthermore, there then needs to be an additional set of physical registers to ensure that all the architectural registers are covered. In such a case, we may view the architectural registers as also being the physical registers and the renaming data table as being an extension of this register set.

Based on the preceding discussion, one possible for an optimized register-renaming system whose organization is as follows. It consists of a data table whose size corresponds to the maximum possible number of outstanding WRITEs, a mapping table of the same size as the data-storage table, and as many physical registers as there are architectural registers. The explicit distinction, made above, between physical and architectural registers now effectively disappears, and renaming data-storage table serves only as a temporary staging post for data values that are eventually written into the registers. Without a one-to-one correspondence between the architectural names and the slots of the mapping table, the mapping table must now be managed in the same way as the cache system: in particular, each slot requires a bit to indicate whether or not it is in use; the need for renaming (i.e. the existence of WAW or WAR hazards) must now be determined by comparing an incoming name with all the valid entries in the mapping table; and a more complex algorithm is required to allocate and free mapping-table entries. A complete system of the type just described is implemented in the PowerPC 604; it is described in Section 5.4.6.

As we saw in Chapter 4, several current high-performance machines deal with the problems posed by conditional branches by making a prediction of the target of the branch and speculatively processing instructions in the predicted path until the target of the branch is definitely known. Where such speculative processing includes the execution of the chosen instructions, computed results cannot be committed to their final destination until it has been ascertained that the instructions were from the correct path. To facilitate rolling back the execution, it is usual in such cases to initially store the results in a temporary buffer and then either write them to their destinations or discard them, according to the definite outcome of the branch-instruction execution. One such buffer is commonly referred to as a *reorder buffer*, as its use must also ensure that, even with out-of-order processing, the results are committed in the order indicated by the original program sequence. (Reorder buffers are, as we shall see in Chapter 7, useful for exception-handling in general.) Where a reorder buffer is implemented, it can also be used as a renaming table.

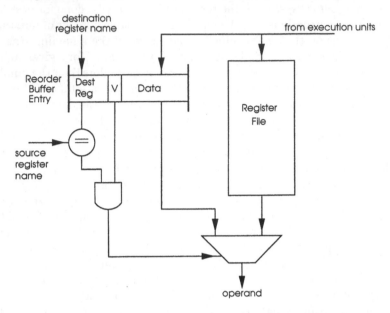

Figure 5.5.   Data-hazard use of a reorder buffer

Figure 5.5 shows an entry in a typical reorder buffer and its use; many practical implementations tend to be just minor variants of this. The structure and use of such a buffer is as follows. (We shall assume a register-architecture in what follows.) The allocation and de-allocation of entries in the buffer is in a first-in/first-out order, as the buffer must track the instruction-issue order if results are to be committed in that order, and an entry for a given instruction is deleted once the results of the instruction have been committed to the register file. When an instruction is dispatched, it is assigned the next free entry in the reorder buffer. The destination-register name is then copied from the instruction into the Destination-Register field of the allocated line, the Validity (V) bit of the line is reset (to indicate the absence of valid data), and the address of the allocated line (or some other unique identifier) is appended to the instruction. (Since each decoded instruction is assigned a new line, renaming is simply accomplished.) Subsequently, when the instruction has been executed, the correct line in the reorder buffer is identified, the result is written into the Data field, and the Validity bit is set. RAW dependences are detected by associatively comparing each source-operand-register name with all the names in the Destination-Register field of the buffer. Because of the manner in which the renaming is carried out, there can be several entries that match in this comparison; evidently, only the most recent entry is the correct one, and it can be identified by implementing an "age-priority" algorithm or by using something like the cache's mapping bit above. If the validity bit on the selected line is set, then the data is read out; if the bit is not set, then, again, the address of the line (or some other unique identifier) is appended to the instruction, which then waits for the data to become available. And if there is no matching entry, then it means the result has already been written into the register file and the original reorder-buffer entry deleted; the data is therefore obtained from the register file. Thus the data table of Figure 5.2 is now split between the reorder buffer and the register file. A system of the type just described is implemented in the AMD K5 microprocessor (Section 5.4.7).

5.3 FAST RESOLUTION OF TRUE HAZARDS

The execution of a program can also be speeded up during the resolution of hazards on memory locations by "short-circuiting" the steps involved in reading and writing data between memory and registers; this is especially useful for memory operations, as these can be of lengthy durations (relative to register and computational operations). The procedures here are known as (*internal*) *forwarding* and *overwriting*, and there are three types: *load–store forwarding*, *load–load forwarding*, and *store–store overwriting*. Load-store forwarding involves the case where there is a RAW hazard on a memory location whose contents are written into a register; that is, a memory location is written into by one instruction and then read from by another instruction while the data is still available within some processor register. In this case, the processing of the sequence of operations can be speeded up by eliminating the second memory access. Load-load forwarding involves the case where the contents of one memory location is read, by two different instructions, into two different registers; here, the second memory access can be replaced by a register-transfer operation. Store-store overwriting is the resolution of a WAW hazard on a memory location by effectively eliminating the first WRITE; actual elimination may not be desirable if the sequential-order semantics of the program are to be maintained, but the desired effect may be obtained in other ways. Examples of all these cases are given in Table 5.3 (in which M denotes a given memory location).

The example of Table 5.3(a) shows how to speed up the resolution of a RAW hazard by bypassing the reading from memory; accordingly, the technique is known as *bypassing*. The same idea can be extended to other types of RAW hazard — in particular, to the case where Ri above is replaced with a more general expression; an example is shown in the first column of Table 5.4. A further extension use the same technique in cases where only registers and computational instructions are involved. As an example, the code fragment of Table 5.1(a) would be transformed as shown in the second column of Table 5.4. It might appear that the transformations of Tables 5.3 and 5.4 could have been carried out by software. This, however, may not be desirable or even possible: the replacement indicated in Table 5.4 would, if carried out in software, imply a repetition of the arithmetic operations, and in the examples of Table 5.3, it is not even possible for software to always carry out such transformations, as, in general, the memory addresses might be computed at run-time, in which case

Table 5.3.   Forwarding and overwriting

Load-Store	Load-Load	Store-Store
M ⇐ Ri	Ri ⇐ M	M ⇐ Ri
Rj ⇐ M	Rj ⇐ M	M ⇐ Rj
effectively	*effectively*	*effectively*
becomes	*becomes*	*becomes*
M ⇐ Ri	Ri ⇐ M	
Rj ⇐ Ri	Rj ⇐ Ri	M ⇐ Rj
(a)	(b)	(c)

a compiler cannot always determine whether or not the same address appears in both instructions. What is intended in all these examples is the indication that when the hardware detects the situations shown (in the originals), it will process the second instruction as though the code were that given by the transformation. We next discuss how this can be done.

Table 5.4.   Bypassing

Case 1	Case 2
M ⇐ Ri+Rj	Ri ⇐ Rj*Rk
Rk ⇐ M	Rl ⇐ Ri+Rm
effectively	*effectively*
becomes	*becomes*
M ⇐ Ri+Rj	Ri ⇐ Rj*Rk
Rk ⇐ Ri+Rj	Rl ⇐ Rj*Rk+Rm

The transformations above (Table 5.3) that involve memory are easily realized through the use of Load and Store Memory Buffers (Chapter 3). Let us assume that in each case the data for the first instruction has already been computed. Then in each case, if the first instruction has not been completed, there will be a corresponding entry (consisting of an address and data) in the appropriate buffer, and the conflict indicated can be detected by looking for a match (via associative comparison) between the address from the second instruction and all the addresses in the buffer. Thus in the first case the transformation means that Rj is loaded directly from the Store Buffer; in the second case it means that Rj is loaded from Ri or that Rj is loaded from the Load Buffer (possibly concurrently with the loading of Ri); and in the third case, it means that the result of the first LOAD is marked in such a way that following instructions see only the effect of the second, with any intervening accesses being satisfied from, say, the Load Buffer. If the data in question is not already available, then in the first case there will be an entry in the Load Buffer and one in the Store Buffer, and when the data becomes available both M and Rj can be loaded at the same time. The details of the logic required for the implementations are discussed in Chapter 3.

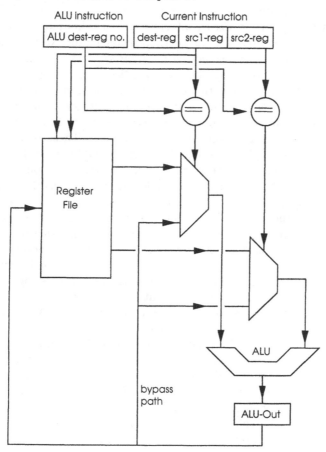

Figure 5.6.   Implementation of bypassing

The examples of Table 5.4 require an implementation in which data can be passed directly between the output and the input of an execution unit. Let us consider a simple implementation, based on the code fragment of Table 5.1(a). In normal processing, the second instruction has to wait for the first to complete, resulting in a delay that includes the time required to write into R0. Suppose, however, that the multiplication is completed just after the reading of R3. Then, instead of waiting for the writing of R0 to be completed, the ADD instruction could be immediately forwarded to the execution unit, with the result of the multiplication also being routed back to the unit at the same time; the writing into R0 is then overlapped with the addition. This is shown in Figure 5.6 and is known as *bypassing*, *short-circuiting*, or *data forwarding*. The separation of instructions involved in bypassing may be greater than one, but more hardware (ALU output registers, multiplexors, and control logic) is then required. It should now be apparent that the logic of Figure 5.1 implements simple versions of such optimizations: forwarding and the supression of an earlier write operation.

The structure shown in Figure 5.6 is nevertheless limited in two main ways: the first is that it requires that instructions be issued for execution in program-order, even though some instruction following one involved in a dependency might still be able to proceed; the second is that it is not easily extendible to more than one waiting instruction or one functional unit, as the logic required to detect the dependency between an incoming instruction and an instruction already in execution and to properly sequence the instructions rapidly becomes complex with more instructions or functional units. These limitations can be dealt with as follows. (Assume that there are several functional units.) "Waiting stations" are placed between the immediate pre-execution stages and the execution stages. The waiting stations hold any instructions that are stalled on a hazard, but allow other instructions that are not affected to proceed, provided they can do so without in the end affecting consistency (i.e. the sequential-order semantics of the program). Given that a number of instructions will now be waiting for operands that eventually come from different functional units, and in an arbitrary order, some means is required to associate the instructions with the data. This association can be done by assigning a unique tag to each result and the same tag to any instruction that will use the result. (The tag may, for example, be the address of a line in a renaming table.) When the result is available, the waiting stations are searched for an instruction with a tag that matches the result-tag, and any such instruction that is found is marked to indicate that that particular operand is now available; instructions that have all operands ready are issued for execution in some predetermined order (e.g. by the order of arrival at the waiting stations). Since results may also have to be written to registers or memory locations, the tag on a result is is also assigned to a register or slot in the memory Store Buffer as necessary; and whenever a result is available, both registers and the Store Buffer are searched in the same way as the waiting stations and the result read in if there is a match on tags.

The above arrangement allows several instructions to wait for one result and, consequently, for several RAW hazards to be resolved in one step when the awaited result becomes available. There are two well known types of implementation of the system just described. The first includes the use of renaming to remove artificial dependencies, and thus increase the independence of instructions in the waiting stations, and permits data to be stored in the waiting stations themselves, thus eliminating the need to read from the register file once the data is available. This implementation is known as *Tomasulo's algorithm*, after its inventor, and was first implemented in the IBM 360/91 (which is is described below); the waiting stations have come to be known as *reservation stations*, after the IBM 360 terminology.

In a machine that implements reservation stations it is possible to arrange it so that the data table of Figure 5.2 is essentially embedded in the reservation stations (and the physical registers that correspond to the architectural registers) and for the mapping table to be implicit in the mechanism that allocates the reservation stations. In the second type of implementation, there is not necessarily any renaming and data is not stored in the waiting stations: results are stored only in registers and are read from there for instructions that are marked as ready for execution. This type of system is commonly referred to as a *scoreboard*, after the terminology used in the CDC 6600 (Section 5.4.2), in which it was first implemented; a system with some similarities to the CDC 6600 scoreboard is implemented in the the VAX 8600 and in the HP PA-8000 (a recent implementation of the HP Precision Architecture).[2] In addition to differences already mentioned above, a further difference between the reservation stations and the scoreboard is that the reservation stations can be centralized or distributed among the functional units, whereas the scoreboard must necessarily be centralized. It is possible to implement something in between the two, and the MIPS R10000, for example, has a system that is similar to a scoreboard but with renaming. Most current high-performance machines implement a variant of Tomasulo's Algorithm.

The idea of reservations stations has been pushed to its logical conclusions in *dataflow* (or *data-driven*) computers in which an entire program is nothing more than a large data-dependence graph, and the entire program memory functions as one very large set of reservation stations.

5.4 CASE STUDIES

In this section we shall look at some implementations of the techniques discussed above. The first case study is the Manchester MU5 primary pipeline; this shows a simple approach to dealing with hazards: the pipeline is designed so that of the three types of hazard, one cannot occur, one requires stalling, and the third is handled in a non-optimal way [20]. The second case study is the CDC 6600; this is an example of scoreboarding and is the first implementation of that technique [28]. The third is the IBM 360/91; this was the first implementation or renaming with forwarding, and all current implementations of such an arrangement are, essentially, derived from that implementation [29]. The fourth case study is the IBM RS/6000; this is an example of renaming with an explicit mapping table, as described in Section 5.2 [15]. The fifth case study is the MIPS R10000 implementation [19]; this is also an example of explicit-map renaming, but in a system that is similar to a scoreboard and which is used in conjunction with completion reordering. The sixth case study is the PowerPC 604, which is an example of the optimized register renaming [27]. The seventh

---

[2]The reader should be aware that the term *scoreboard* is nowadays used by some to refer to *any* hardware mechanism for the detection and resolution of hazards.

is the AMD K5; this shows how renaming can be combined with the use of a reorder buffer. And the last example is the Metaflow Lightning, which shows how renaming, reordering, and reservation-station functions can be integrated within a single unit [22].

5.4.1 MU5 primary pipeline: simple hazard detection and resolution

The MU5 Primary Pipeline exemplifies a simple approach to the detection and resolution of hazards. The pipeline is shown in Figure 1.5 and has been partially described in preceding chapters. Since all named variables are read from and written into the Name Store (a small fully associative cache, described in Section 3.2), this is the point at which hazards must be detected and resolved. A RAW hazard occurs when an instruction needs to read a line of the Name Store, but that line is due to be written into by an instruction whose execution is yet to be completed by the secondary arithmetic unit (B unit). In order to be able to detect such a situation, when an instruction that writes from the B unit into the Name Store arrives at that stage of the pipeline, and there is no uncompleted WRITE instruction at the B unit, the number of the line it will write into is recorded in the B-Write (BW) register and the instruction then proceeds to the B unit. A RAW hazard is then detected if a subsequent instruction tries to read from the line addressed by the BW register. The resolution is to stall the pipeline until the awaited value is returned from the B unit.

A WAR hazard cannot occur in the pipeline, as the pipeline issues and executes executes all instructions in their initiation order, and, therefore, any line of the Name Store will have been read from before there is an attempt to write into it. (Similarly, in the Intel Pentium, the WAR hazard cannot occur because instructions are issued after reading, but the writing takes place at a later stage than the reading does [2].) The case of a WAW hazard is slightly more complex: since there is only one BW register, there can be at most one outsanding WRITE; that is, the pipeline cannot handle two WRITEs, even if they are to different lines of the Name Store. The pipeline therefore detects a *pseudo WAW* hazard if one WRITE instruction arrives at the Name Store while there is another WRITE outstanding and resolves this by stalling the pipeline until the earlier WRITE is complete.

There are a number of more or less obvious extensions that could be made to the arrangement just described: for example, genuine WAW hazards could be catered for by having several BW registers, renaming could be added as discussed above for caches, and so forth.

5.4.2 CDC 6600: scoreboarding

The CDC 6600 machine is at the beginning of superscalar processor design: it consists of a fairly short instruction pipeline, with performance being derived from multiple arithmetic function units that can operate in parallel. A high-level organization of the machine is shown in Figure 5.7. Instructions enter the processor through the Instruction Stack (a loop-catching buffer described in Section 4.3) and are dispatched to the functional units by the *scoreboard*. Each functional unit takes its operands from the twenty-four scratchpad registers — eight 60-bit operand (X) registers, eight 18-bit address (A) registers, and eight 18-bit index (B) registers — and returns results to these registers; the function of the Scoreboard is therefore to detect and resolve dependencies on these registers. Instructions may be executed out-of-order and may be dispatched before their operands are available; the Scoreboard carries out the sequencing that is required to ensure correctness.

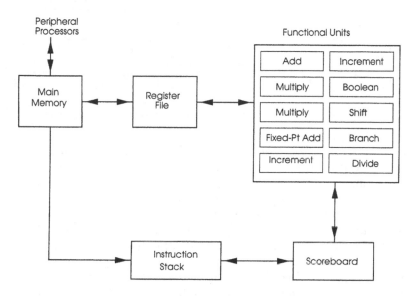

Figure 5.7.   CDC 6600 central processor

The mechanism used by the Scoreboard centers around *reservations* that are placed on registers at the time of instruction dispatch, *flags* to indicate the availability of operands, and *function designators*, which are numbers that uniquely identify functional units. Figure 5.8 shows these flags and designators for a functional unit that takes two input operands and produces one output result. The F registers identify the source registers (Fj and Fk) and the destination register (Fi) in an instruction, the Q registers (Qj and Qk) identify the functional units that will produce the two source operands (if the instruction is dispatched before the operands are available) and so link producers and consumers of results, and the Read Flags indicate the availability of the two operands. So making an entry in an F register for which the corresponding flag indicates the absence of operand-data is, in effect, a reservation to read from that register. Each register also has an associated *result identifier* that, when the register contains no valid data, holds the function designator of the functional unit that will write into it; if the register contains valid data, then its Result Identifier has the value zero. So, in effect, the Result Identifier indicates whether or not a register has been reserved for writing and by which functional unit if it has been.

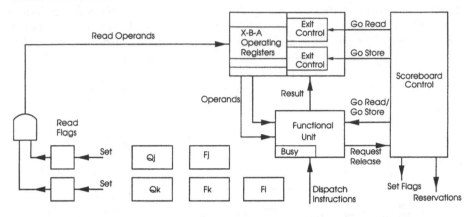

Figure 5.8.   CDC 6600 functional unit reservation designators

The operation of the above arrangement is as follows. Assuming that there is an appropriate functional unit available (i.e. not busy), the Scoreboard dispatches an instruction after setting the F registers from the operand-specifiers in the instruction, and, for operands that are not available, setting the Q registers from the Result Identifiers of the source-operand registers, marking the functional unit as "busy", and entering the number of the functional unit allocated into the Result Identifier of the destination register. Once an instruction has been dispatched, it reads from its operand-registers and is executed only if the Read Flags are set (which corresponds to Q registers with values zero), and it may store its result only when the Scoreboard control issues it with a signal indicating that it may do so. (The system is designed so that, as far as possible, several functional units can concurrently read from or write to the register file.) On the completion of an instruction, a function unit, under the direction of the Scoreboard, writes into its destination register, releases the reservation on that register, and broadcasts its (the functional unit's) designator to all the other functional units. The latter units will then compare the broadcast designator with those in their Q registers and proceed to set the corresponding Read Flag for any waiting instruction if the comparison results in a match. When both Read Flags are set for a waiting instruction, the operands are read from the registers specified by the F registers, the Read Flags are reset, and the instruction is then executed. The Scoreboard issues the signals that control these phases — *go-read* (to initiate the reading), *go-store* (to initiate writing), *set-flags*, and *place-reservations*, as needed — and the functional units communicate with the Scoreboard to request permission to read or write.)

A RAW hazard in the CDC 6600 occurs when an instruction requires, as a source operand, the result of a previously dispatched, but as yet uncompleted, instruction; the situation is detected if the source register has been reserved for writing, (as described above). If the functional unit specified in the operation code of the reader instruction is available, then the instruction is dispatched, but the READ is held up in the functional unit until the WRITE is completed and the reservation on the register of conflict has been removed. There may, in fact, be a number of READs pending on a single register, and the removal of the reservation will allow all of them to proceed. The WAR hazard occurs when an instruction needs to store its result in a register whose contents are still to be read by a previously dispatched but as yet unstarted instruction. (Recall that an instruction cannot be started before the operands have been read.) In this case the WRITE instruction is dispatched, and may executed, but the actual writing of the result is held up until the reservation placed on behalf of the READ has been removed; the stall on the writing of the result is therefore postponed to the last possible moment. (The issuing of the Go-Store release signal for the WRITE is complicated by the fact that the only way to determine whether any instruction may be waiting to read from a register that is about to be written into is through the Read Flags and F registers: an operand register has a waiting reader if that register is named in an F register and the corresponding Read Flag is set. Furthermore, there may be several waiting readers. So the logic to determine when a Go-Store signal may be issued appropriately combines information from all F registers and Read Flags.) And the WAW hazard occurs when the destination register of one instruction is also the destination register of a previously dispatched but as yet uncompleted instruction. This hazard is resolved by not dispatching the second WRITE until the first is complete and the reservation on the destination register is removed. The handling of the WAW hazard therefore introduces a delay that is greater than that for a WAR hazard, but it should be noted that a first-order WAW hazard is a very improbable event that is likely to arise only as a result of a programming error, and a second-order WAW hazard occurs only as a combination of WAR and RAW hazards.

An obvious extension to the system just described would be to broadcast both data and functional designator when a result becomes available. This would, at the cost of having to provide wider buses, eliminate the reading and associated controls. Such a data forwarding mechanism was first implemented in the IBM 360/91, and variants are currently implemented in several machines.

5.4.3 IBM 360/91: renaming and forwarding

In the CDC 6600, all computational instructions involve only register-register operations, and the problem of detecting and resolving hazards is straightforward. This is in contrast with the IBM 360/91, in which computational instructions can involve register-memory operations; to deal with these, data to and from memory are buffered in intermediate registers, i.e. the instructions are effectively transformed into register-register operations. These intermediate registers and a number of *reservation stations*, and their use, in fact constitute a limited renaming system in which they serve as physical registers.

The organization of the IBM 360/91 floating-point unit is shown in Figure 5.9. Instructions are dispatched through the Floating Point Operand Stack

(FLOS), which decodes and routes them to the appropriate functional unit. The FLOS maps both memory-register and register-register instructions into a pseudo register-register format in which one operand-name is one of the four Floating-Point Registers and the other is a Floating-Point Register, a Floating-Point Buffer (for operands received from memory), or a Store Data-Buffer (for operands being written to memory).

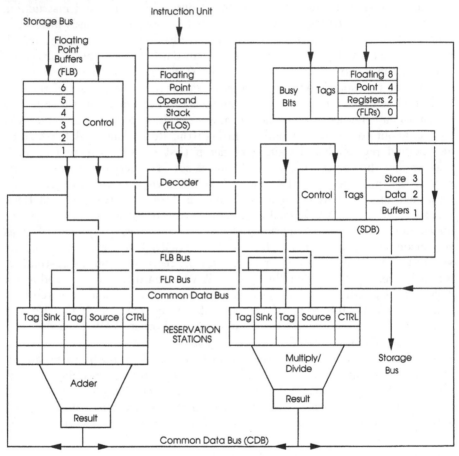

Figure 5.9.   IBM 360/91 floating-point unit

There are three major features in the design of the floating-point unit: The first is the Common Data Bus, which is fed by all units that can alter a register and which feeds the Floating-Point Registers, the Store Data-Buffers, and all units that can have a register as an input operand. (The latter connections allow data produced as the result of any operation to be forwarded to the next execution unit without first going through a Floating-Point Register, thus reducing the effective pipeline length for a RAW hazard.) The second significant feature is the reservation stations at each functional unit to hold instructions that have been dispatched but are held up on some hazard; each station consists of some slots for the source operands, the identity of the units supplying the source operands, and some control information. The third significant feature is the system of *tags* employed to realize forwarding and partial renaming. A tag is a four-bit number generated by the Common Data Bus control logic to uniquely identify each of the eleven sources that feed the Common Data Bus: the six floating-point buffers and the five reservation stations. Tag registers are associated with each of the four Floating-Point Registers and with each of the four Store Data-Buffers. Each Floating-Point Register also has a *busy bit* that is set whenever the FLOS dispatches an instruction with the register as a destination and reset when the register is written into. (The Busy bit therefore has a function that is similar to that of the Result Identifier in the CDC 6600.)

The arrangement just described operates as follows. An instruction is dispatched only if there is a reservation station available at the appropriate functional unit. When the instruction is dispatched, the FLOS sets the Busy bit on the destination register, enters a new tag in the tag field of the same register, and attaches the tag to the instruction; so the tag always identifies the last source of a result for that register. Whenever the FLOS decodes an instruction, it checks the Busy bit of each of the floating-point registers named in the instruction. If the register is an operand source and the Busy bit is not set — that is, there is no pending WRITE on the register — then the contents of the register are sent, via the Floating-Point-Register Bus, to the assigned reservation station. If the Busy bit is set, then the tag field of the register will hold the tag of Writer (i.e. a reservation station or Floating-Point Bufer), and this tag will be sent to the reservation station, in place of the register contents. Similarly, when the operand source is a Floating-Point Buffer, either the required data or the tag associated with the register is sent to a reservation station, according to whether or not the data has arrived from memory. Whenever a result appears on the Common Data Bus — from the Floating-Point Buffers or from a functional unit that has completed the execution of an instruction — both the result and the corresponding tag are broadcast to all destinations on the bus. Each reservation station or Store Data Buffer that is waiting for an operand compares its tags with the broadcast tag and, if there is a match, reads in the corresponding data and sets a control bit to indicate the availability of the data. At the same time, a comparison is carried out with the tag fields of all busy Floating-Point Registers, and if there is a match, the result is read in from the Common Data Bus and the corresponding Busy bit is reset. An instruction may be dispatched for execution as soon as both its operands are available at its reservation station.

With the system above, a RAW hazard occurs when an instruction finds the Busy bit set on a source register (by a previously dispatched but as yet uncompleted instruction). The READ instruction is tagged and forwarded to a reservation station or buffer, thus moving any stalling from dispatch-time to issue-time; the stalled instruction is released when the writer completes and broadcasts the data and a matching tag on the Common Data Bus. A WAW hazard occurs when an instruction finds the Busy bit set on its result destination register. In this case renaming (with a tag as the new name) takes place and the second WRITE is also dispatched. The first WRITE will not alter the contents of the register, since the tag on its result will not match the tag (which is that of the second WRITE) that is on the register, and all intervening READS will have had the source operand-names replaced with the tag from the first WRITE; the result in the register will therefore reflect the second WRITE, and the hazard is essentially eliminated. (Observe that this realizes an optimization similar to that in Table 5.3(c).) The WAR hazard is also similarly eliminated: a READ either finds the Busy bit not set and proceeds with the data, or it proceeds with just a tag; the WRITE is not stalled, since a tagged READ waits at a reservation station and therefore causes no conflict on the register.

To relate the arrangement just described to Figure 5.2, we observe that the combination of buffers, reservation stations, and (sometimes) registers serve as the data table (i.e. physical registers); and the register file (sometimes) serves as the mapping table. The "sometimes" accounts for the fact that a register-file entry is a mapping-table entry when the register has no valid data but is a data-table entry when there is data. The system described is easily extendible to include more functional units, reservation stations, buffers, registers, etc.

5.4.4 IBM RS/6000: explicit-map renaming

The IBM RS/6000 is an example of a recent superscalar microprocessor and is the precursor to implementations of the Power architecture. The renaming system is essentially an implementation of the basic system described at the beginning of Section 5.1. Because the machine has only two arithmetic functional units (fixed-point and floating point), each with short (two-stage) pipeline, the detection and resolution of hazards is quite straightforward.

The renaming system is used for only floating-point operations (arithmetic operations and memory LOADS and STORES of floating-point data) and has the high-level organization shown in Figure 5.10; the physical registers (i.e. data table) are not shown. The architecture specifies 32 registers, but the implementation has 40 physical registers. The Map Table therefore consists of 32 entries, each of which is six bits wide, to map the architectural registers to the physical ones. The Free List keeps a record of the (up to eight) physical registers that that at any given moment are not associated with any physical register; initially, architectural register i is mapped to physical register i, and the Free List contains the addresses of the remaining registers. The list is maintained as a circular queue, with register addresses entered in the order in which physical registers become free. The Decode register holds the next instruction to be issued for execution, and the Instruction Decode Buffer (IDB) holds instructions that have been renamed but cannot yet enter Decode because that stage is busy. The Pending-Target Return Queue (PTRQ) is an 8-entry circular queue that contains the addresses of the physical registers associated with the destinations of the instructions in the Decode or IDB registers. Once an instruction passes through the Decode stage, its associated register names are moved from the PTRQ to the Free List; a Release Pointer indicates the next entries to be freed.

The Pending-Store Queue (PSQ) holds the source physical-register addresses of all STOREs that are awaiting data, and the Outstanding-Load Queue (OLQ) holds the physical-register address of the next LOAD instruction that will receive data from the cache/memory. When an instruction enters Decode, its source-register addresses are compared with the contents of the OLQ, and if there is a match, then a RAW hazard exists, and the instruction is temporarily stalled. Each entry in the IDB and Decode registers has two extra fields, Load Count and Store Count, that are used to determine when registers can be released to the Free List: the Load Count is used to increment the Release Pointer for the PTRQ, and the Store Count is used to increment the Release Pointer for the PSQ. (The RS/6000 is one of the very few machines that employs counters in this explicit manner.)

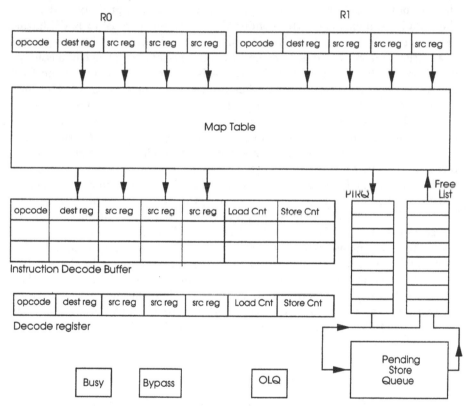

Figure 5.10.   Renaming mechanism in the IBM RS/6000

The Busy register holds the destination-register address of instruction in the first execution stage of the arithmetic unit, and the Bypass register holds a similar address for the instruction in the other stage. As each instruction enters Decode, its source-register addresses are compared with the contents of the Busy and Bypass registers; a match indicates a RAW hazard. If the match is on the Bypass register, then the instruction is allowed to proceed but receives its data through a bypass path (instead of from the register file); and if the match is on the Busy register, then the instruction is stalled, as the instruction on whose result it depends has not proceeded far enough for a data-bypass.

The processing of the different types of instruction is as follows. For a STORE instruction, the architectural source register is renamed to the previously allocated physical register that (eventually) holds the data, and the 5-bit architectural-register address is replaced with the 6-bit address of the physical register. The address of that physical register is then placed on the PSQ until the writing operation is completed. Since the latest arithmetic instruction could be the source of the data to be stored, the Store Count for that instruction is incremented; this stops the STORE from being carried out until that arithmetic operation has been completed. The STORE is finally carried out when it is at the end of the PSQ (since STOREs are done in sequential program-order) and the data is available; the latter condition holds when there is no RAW hazard between the STORE and the source instruction (in execution) or a memory-LOAD instruction underway and is detected by an absence of equivalence in a comparison between the names source physical-register and the contents of the Busy and OLQ registers. Once the data has been stored, the address of the source is removed from the PSQ and returned to the Free List. A LOAD instruction is processed by first renaming the destination register (by assigning a free physical register and updating the Map table), entering in the PTRQ the address of the last physical register associated with that architectural destination, entering in the OLQ the newly allocated physical register, and incrementing the Load Count of the latest arithmetic instruction. For an arithmetic instruction, source and destination registers are renamed, with a new (free) physical register for the destination, the Load Count and Store Count are initialized, and the instruction sent to the Decode or the Instruction Decode Buffers. (Instructions enter the renaming assembly through the registers R0 and R1, on an alternate basis, and the counts are initialized to 0 if the instruction is in R1 and to 1 if it is in R0 and there is a LOAD or STORE instruction in R1.) Once there is no RAW hazard (see above) that requires stalling, the instruction is issued for execution, and unneeded registers afterwards released to the Free List: the instruction's Store Count is used to increment the Release Pointer for the PSQ and the Load Count that of PTRQ.

5.4.5 MIPS R10000: explicit-map renaming

The MIPS R10000, a recent implementation of the MIPS IV architecture exemplifies the use of basic explicit-map renaming, with more physical registers than architectural registers. Unlike the IBM RS/6000, however, the renaming table is also used in conjunction with completion reordering; the overall implementation also has some similarities with a CDC 6600-type scoreboard.

The renaming table consists of 64 physical registers for 32 architectural floating-point registers, 64 physical registers for 33 architectural integer registers, 32-line and 33-line mapping tables, and a list to keep track of free physical registers. Each physical register also has an associated *busy bit*, in a separate table, to indicate whether or not the register is awaiting data. The machine has out-of-order instruction execution, and up to 32 instructions can be in progress at any given time; but, by freeing physical registers in an appropriate order, it can be ensured that precise state can be restored in the event of an exception.

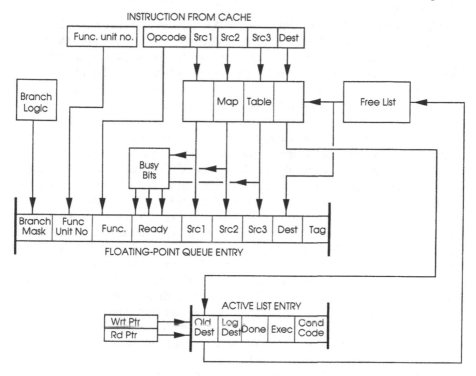

Figure 5.11.   Renaming mechanism in the MIPS R10000

The basic renaming system is quite straightforward: the floating-point part is shown in Figure 5.11, and the integer part is mostly similar. Provided neither the list that keeps track of instructions in progress nor the list of instructions awaiting operands is full — if either is full, then the pipeline is stalled until the situation changes — up to four instructions are decoded in each cycle. For each instruction the architectural register named as the destination is mapped to a new physical register, and the corresponding Busy bit is set; at the same time the architectural source-register addresses are mapped to previously as-signed physical-register addresses in order to get the source operands from the right locations. In order to facilitate a precise rollback in the event of an in-terrupt, the Active List, a 32-entry first-in/first-out (except as specified below) queue, contains for each instruction the address of the architectural destination-register, the address of the physical register (if any) that holds the last value for that architectural destination-register, and some status information. When an instruction completes successfully, its entry in the Active List is deleted and the physical register name is entered in the list of free registers. (Thus if an exception occurs, the Active List can be used to restore the last values of registers.) Since up to four instructions can complete in any cycle, and an architectural register might have been renamed more than once within the group, the deletion/freeing is done in last-in/first-out order for each completing group of instructions.

The renaming eliminates WAW and WAR hazards, leaving only RAW hazards, which are detected and resolved at two levels in the system: the first is between the (up to) four instructions that are dispatched in a given cycle, and the second is between instructions that are dispatched in different cycles. In the first case, twenty-four 5-bit comparators are used on the physical register names to detect equivalences between source-register names and destination-register names; where there is equivalence, the outputs of these comparators control multiplexers that realize bypass paths and replace some register names with names from the list of free registers. In the second case, the instruction is entered in one of three random-order "queues" (associated with the integer execution units, the floating-point execution units, and memory addressing). Each of these queues, which serve as waiting stations, has sixteen lines; and each line has, for one instruction, a four-bit number that indicates what pending (speculative) branches that the instruction depends on (which number is used to determine whether or not the instruction should be aborted after a branch is finally executed), four bits that indicate what functional unit should execute the instruction, a 10-bit function code, three physical-register addresses for the source operands and three Ready bits to indicate the availability of the operands, a physical-register address for the destination, and a tag that is the address of the instruction's entry in the Active List. Whenever a result is produced by a functional unit, the destination-register address on the result is compared with the source-register addresses of all waiting instructions, and the Busy bit for the named physical register reset if there is a match. When an instruction has all of its operands available (i.e. all Busy bits are reset and the Ready bits set), then the instruction is removed from the queue, the operands are read, and the instruction issued for execution. The order in which instruction are removed from the queue is by address (rather than by age) in order to keep simple the necessary logic.

5.4.6 PowerPC 604: optimized register-renaming

The PowerPC 604 (Section 1.5.4) implementation is of a superscalar pipeline in which instructions can be processed out-of-order. The hazard detection and resolution mechanism used includes register renaming. (The registers that make up the renaming structure are also used to facilitate in-order completion of instruction processing; this latter aspect is discussed in Chapter 7.) The complete system also includes reservation stations at each functional unit, and these are managed in a manner similar to those in the IBM 360/91 (Section 5.4.3).

The renaming structure in the PowerPC consists of a Rename Buffers (that corresponds to the mapping table in Figure 5.2) for each set of architectural registers where hazards can arise. Thus there is one buffer for the integer (general-purpose registers), one for the floating-point registers, and one for the condition-code register. Both of the computational Rename Buffers have fewer entries than the number of corresponding registers and are used in the optimized manner described in Section 5.2: the Integer Rename Buffer has 12 registers, which are used for renaming with operations involving the 32 integer registers, the Floating-Point Rename Buffer has 8 slots for use with 32 architectural registers, and there is a separate physical register for each architectural register. The Condition Rename Buffer, on the other hand, has eight slots for use with four conditions. The format of the rename buffer entries is similar in all the buffers, except for differences in the field that holds data; the use and management of the buffers is also similar.

Rename Valid	Register Number	Result	Result Valid

Figure 5.12.   Format of PowerPC 604 rename-buffer entry

Each register in a rename buffer has the format shown in Figure 5.12: the Rename Valid field indicates whether or not a register is in use, the entries in the Register-Number field serve as the Mapping Table of Figure 5.2, the entries in the Result field correspond to part of the Physical Registers in Figure 5.2 and temporarily hold the eventual results, (the register file forms the rest of the Physical Registers), and the Result Valid field indicates whether or not the result has been written in. When the Dispatch stage of the pipeline (see Figure 1.9) detects an instruction that writes into a register, it allocates a new entry in the appropriate Rename Buffer, writes the destination-register address specified in the instruction into the corresponding Register Number field, sets the Rename Valid bit, and resets the Result Valid bit. The destination-register address in the instruction is then replaced with the number of the Rename Buffer register, and the instruction is dispatched. Subsequently, when the instruction executes, it writes into the Result field and sets the Result Valid bit. After the instruction has been processed in the Completion stage of the pipeline (and it has been ascertained that there is no exception that requires a rollback), the result is copied into the register specified in the Register Number field and the Rename Buffer line is freed (by resetting the Rename Valid bit). The order in which data is moved from the Rename Buffer into the register file is determined by a separate reorder buffer.

Using this structure, the locations (buffer or register file) of data values are determined and RAW hazards detected as follows. Prior to allocating an entry in the Rename Buffer, each source-register name given in the instruction is concurrently compared with all active ones in the Register Number field of the Rename Buffer. If there is no match, then valid data already exists in the register file, and this is used for the computation. If there is a match and the corresponding Result Valid bit is set, then the data is read from the Result field. And if there is a match but the Result Valid bit is not set, then a RAW hazard exists; in this case the instruction is dispatched to a reservation station, with the address of the Rename-Buffer line in place of the original operand-address. WAR and WAW are, as usual, eliminated by the renaming.

5.4.7 AMD K5: renaming with a reorder buffer

Data hazards in the AMD K5 are detected and resolved through the use of a renaming table combined with a reorder buffer, as described at the end of Section 5.2. A set of reservations stations, similar to those of the IBM 360/91, is also employed; the whole system is therefore very similar to that of the PowerPC 604, with the only essential difference being in the actual separation of renaming and reordering in latter.

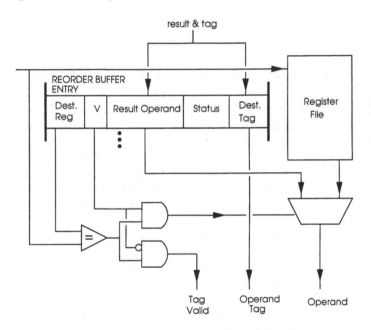

Figure 5.13.   Entry in AMD K5 reorder buffer

Figure 5.13 shows the format of an entry in the reorder buffer and its relationship to the rest of the system. The allocation and de-allocation of entries in the buffer is in a first-in/first-out order. As each instruction is dispatched, it is assigned the next free line in the buffer; this accomplishing the renaming. If the reorder buffer is full, then that part of the pipeline that precedes the renaming stage is stalled until an entry becomes free. An entry is deleted from the buffer, and the result there committed into the register file, when it is at the head of the queue and it has been determined (from the Status bits) that the corresponding instruction completed normally and that it was indeed correct to execute it (i.e. that if there was a branch instruction ahead, then the instruction was on the target path). If an entry reaches the head of the queue before the corresponding result has been returned, then retirement is stalled, but allocation of new entries continues as long as there is space in the buffer.

Once a line in the buffer has been allocated, the destination-register name is copied from the instruction into the Destination-Register field of the allocated line, the Validity (V) bit of the line is reset, and a unique Destination Tag associated with the line is appended to the the instruction. When the instruction has been executed, the tag on the result is associatively compared with the contents of the Destination-Tag field in all the lines of the buffer and the data is written into the line that yields a match; the Validity bit of any matching line is then set. The renaming eliminates WAR and WAR hazards, leaving only the RAW ones, which are detected by associatively comparing the source-operand-register names with all the names in the Destination-Register field of the buffer for a line with a matching name. (If there there is more than one line with a matching entry, then a priority algorithm locates the most recent one.) Then if the validity bit on the selected line is set, the data is read out; otherwise, the Destination Tag of the line is appended to the instruction. In either case, the instruction is then sent to wait at a reservation station, with Tag-Valid bits set to indicate whether or not there is any accompanying data. In the case that there is no matching entry, the data must already be in the register file and is read from there.

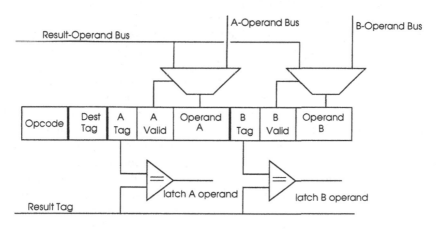

Figure 5.14.   Entry in AMD K5 reservation station

The format of a reservation-station entry is shown in Figure 5.14. The logic associated with the reservation station monitors the result buses for data whose tag matches any of the tags in the reservation station; if there is a match, then the data is read into the corresponding slot and the tag-valid bit set appropriately. (An instruction is ready for execution when both tag-validity bits are set.) Results are also sent to the reorder buffer, where similar actions are carried out.

5.4.8 Metaflow Lightning: combining renaming, reordering, and reservation stations

The hazard detection and resolution system used in the Metaflow Lightning is different from that used in most machines in that a single structure is used for all three of renaming, reordering, and reservation-station functions. The unit used is known as the DRIS (Deferred-scheduling, Register-renaming, Instruction Shelf), and its relationship to the rest of the machine is shown in Figure 5.15. Each entry in the DRIS has the format shown in Figure 5.16; essentially, the Source-Operand sections function as reservation stations, and the Destination part functions as a renaming table. Instructions move from the Dispatch unit into the DRIS and (copies) remain there until the Retire unit commits their results into the register file.

Entries in the DRIS are allocated in program-order. Unless the DRIS is full, in which case the Dispatch unit is stalled until a free entry becomes available, an instruction from the Dispatch unit is allocated the next free slot in the DRIS. A unique identifier (UID) is then assigned to the instruction, which UID the instruction carries through to the execution stage. (The identifiers are addresses of DRIS lines and are issued in strict numerical order, with an additional bit to indicate when wrap-around occurs. The identifiers therefore indicate the "age" of the entries.) Next, the destination-register name is copied from the instruction into the DRIS entry. The Latest bit is then reset for the last entry for that destination-register name and set in the new entry, thus ensuring that it is correctly identified by following READS. (The Latest bit therefore has the same function as the Mapping bit for the cache-renaming system of Section 5.2.) This accomplishes the renaming. The last step in this phase is the writing of source-operand-register numbers from the instruction into the Source Register Number fields of the allocated entries.

WAR and WAW hazards having been eliminated by the renaming, RAW hazards are detected by concurrently comparing the register name for each source operand with the contents of the Destination Register-Number field of all lines. The absence of a match indicates that the operand has already been written into the register file, and it is read from there. Otherwise, i.e. if there is a match, then the matching UIDs are examined to identify the most recent entry. If the data computed by the identified instruction is already available, but has not yet been written into the register file, then it will be in the Content field for the Destination; this status will be indicated by a set Executed bit on the line. Otherwise, the UID of the instruction producing the result is written into the UID field for the required source operand, and the corresponding Locked bit is set. (The Locked bit therefore has the same function as the Tag-Valid/Data-Valid bit in a reservation station and the Read Flag in the CDC 6600 scoreboard.) Subsequently, when an instruction has been executed, the Update unit writes the result into the correct DRIS entry (which is identified by the UID on the result) sets the Executed bit, and (after a concurrent comparison of the incoming UID with those of waiting instructions in the DRIS to find the correct entry) unlocks any waiting operand requests.

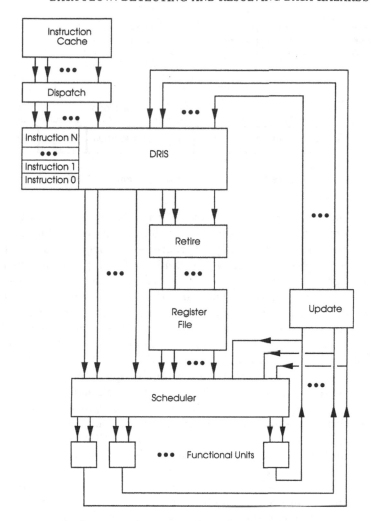

Figure 5.15.   Organization of Metaflow Lightning

The task of the Scheduler is, in every cycle, to determine whether or not there are executable instructions (i.e. those that are not waiting for any operand and for which there is a free functional unit) and to issue for execution the oldest of any such instructions; the issued instruction has its DRIS entry so marked, in order to avoid repeated issue.

5.5 SUMMARY

In this chapter we have discussed the problem of potential discontinuities in the flow of data and its solutions in hardware. These discontinuities arise from dependences (hazards) between instructions in a pipeline and unless resolved in some way, they require that the pipeline be stopped when they occur. The

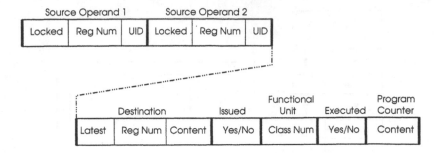

Figure 5.16.   Entry Metaflow Lightning DRIS

basic problem can be solved in a number of ways that require varying degrees of hardware complexity to implement, but in general the best solution (in terms of performance) includes altering program-specified logical names, in such a way that the non-essential hazards are eliminated, and processing instructions out of their program-order (wherever possible). The basic ideas first appeared in the late 1960s, in the CDC 6600 and the IBM 360/91, but at that time they were too complex and costly to implement in most machines. In recent years, however, decreases in hardware costs and the continual demand for high performance have made such techniques more worthwhile, and they now appear in several high-performance machines. With further decreases in hardware costs, we should expect that they will eventually appear in most machines and to an even greater extent in the machines at the high end of the performance spectrum as the the degree to which they are superpipelined or superscalar increases. Other advances may also include the combination of the techniques discussed here with the the partial or full use of multithreading.

This chapter has emphasized only the hardware techniques. Nevertheless, the reader should be aware of the existence of a number of software techniques that can be used in place of the simpler hardware ones discussed here and should also note that even the more sophisticated hardware techniques can be made more effective with the use of proper compiler techniques.

6 VECTOR PIPELINES

For high performance, vector machines rely on the relative ease with which an arithmetic pipeline can be kept busy if the main operations to be carried out are arithmetic operations on vectors and matrices; that is, when there is a large number of elemental operations of the same type and no control-transfers to break the pipeline flow. Vector processors machines are therefore very useful in high-performance scientific computing. For such machines and applications, there is also the additional benefit of low static and dynamic instruction counts, as a single vector instruction generally corresponds to a large number of scalar operations. So, not surprisingly, until recently, the majority of supercomputers have been vector-pipelined, and vector machines (in uniprocessor or multiprocessor systems) still occupy a prominent place at the high end of the performance range. Nevertheless, with current microprocessors performing in what was only recently the "supercomputer" range, and at relatively low costs, the high cost:performance ratios of vector machines render them less attractive than they have been for a long period.

This chapter is an introduction to the design and programming of vector machines. The first section introduces the fundamental concepts, the second section deals with storage-access techniques, the third with instruction sets and formats, the fourth with programming aspects, and the fifth is a discussion of performance issues. The sixth section consists of two case studies, and the last is a summary.

6.1 FUNDAMENTALS

Although most high-performance machines employ pipelining in one form or another in the implementation of arithmetic units, such pipelining in most effective with vector operations. For such operations, it is relatively easy to ensure a steady stream of (scalar) orders of the same type and with a high degree of independence, which is what is required for maximum pipeline throughput. Also, whereas the performance of an instruction pipeline is constrained by the rate at which instructions can be fetched into the pipeline — a limitation known as *Flynn's bottleneck* — this is much less of a problem with vector pipelines, since a single vector instruction encodes a larger number of elemental instructions. Memory interleaving is also more effective with the relatively more regular structure of vector data. All this means that the processing a vector instruction in a vector processor can be much faster than that of an equivalent set of scalar instructions in a scalar processor.

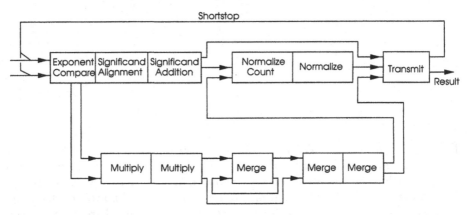

Figure 6.1.   CDC STAR-100 floating-point pipe 1

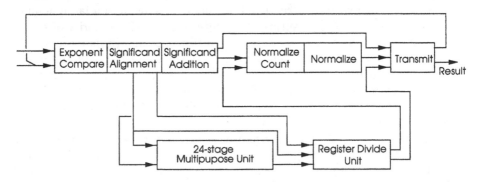

Figure 6.2.   CDC STAR-100 floating-point pipe 2

Arithmetic pipelines have been introduced in Chapter 1, with two examples given from the TI-ASC and the VAX 6000/400. Two additional examples are shown in Figures 6.1 and 6.2; these are the multifunctional pipelines of the CDC STAR-100, one of the first vector supercomputers. The first floating-point pipeline performs arithmetic operations on floating-point operands and fixed-point operations on addresses. The former operations include addition, subtraction, multiplication, comparisons, and so forth; and the latter include indexing operations and operations to manipulate the contents of various registers. The pipeline configuration for addition and subtraction corresponds to the pipeline of Figure 1.2 and consists of six stages: EXPONENT COMPARE, SIGNIFICAND ALIGNMENT, SIGNIFICAND ADDITION, NORMALIZATION COUNT, NORMALIZATION SHIFT, TRANSMIT; the last unit returns the result. Multiplication is carried out in the lower unit, with the result being transmitted directly or first normalized. The multiplication unit is a highly parallel one that splits both operand into two parts, carries out four smaller multiplications in parallel, and then merges the results into a single one. The result of one operation may be returned directly into the pipeline (cf. *bypassing* or *forwarding* in Chapter 5), an arrangement known as *shortstopping* in CDC terminology. The second pipeline is similar to the first, except in the provision of the Register-Divide and Multipurpose units. The first unit carries out all register division operations and conversions to and from binary and binary-coded-decimal; the other unit is a 24-stage pipeline that carries out square-root, vector-divide, and vector-multiply operations. Other examples of arithmetic pipelines are given below with the case studies.

Since a typical program is not comprised solely of vector instructions — some arithmetic instructions are scalar, and other scalar operations are also needed to control the vector operations and for general "housekeeping" tasks — a vector processor must therefore be used in conjunction with a scalar one. Thus the typical organization of a complete vector machine is as shown in Figure 6.3: The main memory is usually highly interleaved. The Instruction Processor typically consists of a short instruction pipeline that fetches and decodes instructions and then issues them to the Scalar Processor or to the Vector Processor, as appropriate; the two processors may operate concurrently. The Vector Memory-Access Controller takes each operand specification (e.g. a base address and vector length), generates to the memory the sequence of addresses needed to fetch the vector elements and sends these addresses to the main memory, and also buffers the operands in the Intermediate Buffer. The arithmetic pipelines take operands from and return their results to the Intermediate Buffer. The connection to main memory is through one or more Load/Store units that may also be pipelined. And the arithmetic pipelines may be unifunctional or multifunctional. The initialization of a vector operation includes fetching and decoding the instructions, initializing indexing counters, and carrying out some non-vector housekeeping operations; the time required to do all this is the vector operation's *set-up time*.

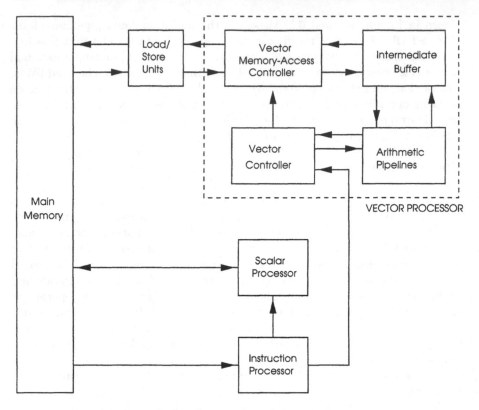

Figure 6.3.   Organization of a vector machine

Register-to-register versus memory-to-memory machines

Vector machines generally fall into one of two broad categories, according to the actual implementation of the Intermediate Buffer of Figure 6.3: In *register-to-register* machines, the buffer consists of program-addressable vector registers; there may, however, be some additional local buffering for instructions that load the vector registers or store their contents to the main memory. In *memory-to-memory* machines, the buffer usually consists a simple programmer-invisible buffer (e.g. a pair of FIFO buffers); arithmetic operations, in principle, take their operands from memory and return results to memory. (A vector cache may be used, but, as discussed below, this is unlikely to be as effective as in a scalar processor.) Most of the current vector machines — those from Cray Research, Fujitsu, NEC, Hitachi, etc. — are register-to-register machines [28, 30, 34, 38]; only Control Data Corporation and its former spin-off ETA have manufactured pure memory-to-memory machines [8, 23]; and the IBM 3090, by allowing a single instruction to have both memory and register operands, aims to get the benefits of both types of architecture [5, 6, 26].

Probably the foremost design issue — one that has far-reaching implications for the implementation, performance, programming, etc. — for a vector machine is whether the architecture is memory-based or register-based. The main factor here is the start-up time: for a memory-to-memory machine this includes the time to access memory and can therefore be very long. Suppose a vector operation is to be performed on vectors of length n and that the machine has a single arithmetic pipeline of k stages, a cycle time of t_p, a memory access time of t_m, and a set-up time of t_s for a vector operation. Then for a memory-to-memory machine, as the operands have to be fetched from memory, the time required for the complete vector operation is approximately

$$T_M \;=\; t_m + t_s + kt_p + (n-1)t_p \tag{6.1}$$

For a register-to-register machine, we may assume that the operands are taken from registers. So if the register length is l, the vectors must be partitioned and processed as $\lceil n/l \rceil$ smaller vectors, a procedure known as *strip-mining*. And if we assume that that the loading and unloading of these registers is overlapped with the execution of useful (computational) instructions, that data in registers is being reused and results are returned to registers, and that setting up each strip-mined vector operation takes a time of $\bar{t}_s$, then the time required is approximately

$$T_R \;=\; t_s + \left\lceil \frac{n}{l} \right\rceil \left[\bar{t}_s + kt_p + (l-1)t_p \right] \tag{6.2}$$

(t_s here is that part of the set-up time that cannot be overlapped with useful operations.)

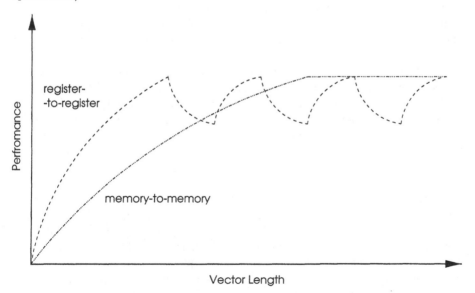

Figure 6.4.   Typical performance of vector machines

Table 6.1.   Vector-pipeline performance vs. vector-length

n	m	Cyber 205	Cray-1
180	16	0.16	0.49
	64	0.40	0.69
	256	0.65	0.69
	1024	0.76	0.69
200	16	0.17	0.53
	64	0.42	0.73
	256	0.66	0.73
	1024	0.78	0.73

Ratios of actual-performance to peak-performance
2-pipeline Cyber; Cray vector-register length $= 64$

Evidently, for typical parameters, the start-up time in T_M will be much larger that in T_R. (Figure 6.4 shows performance curves comparing the two types of vector machine according to the performance equations above, and Table 6.1 gives some figures from actual measurements on an $n \times m$ Fast Fourier Transform [36].) All else being equal, the memory-to-memory machine takes longer to reach peak performance, but the register-to-register machine may shown slight performance drops with every strip-mined piece of a vector. This gives a strong argument for a register-to-register machine, as it is more likely to give good performance on both short vectors and long vectors. Furthermore, the need to access memory for all operands means that that the memory-to-memory machine will require an extremely high-bandwidth memory. As with scalar machines, the vector pipelines can benefit from the use of a cache, but the extent to which this is so is much smaller: the large size of the datasets involved, the comparatively low temporal locality in the usage of these data sets, and the (sometimes) random access to substructures of vectors all reduce the effectiveness. Vector caches are therefore not as useful as caches in scalar machines and have rarely been implemented; one exception to this is the IBM 3090, an implementation of an extended IBM System/370 architecture, in which a cache that is also used for vectors supplements a set of registers [5, 6, 26]. (Other studies of vector caches may be found in [35, 40].) Nevertheless, the register-to-register architecture does have two drawbacks: One is that the programmer or compiler must carry out the required partitioning of long vectors, and, in general, the fixed number and length of the vector registers can be restrictive. The other drawback is in the overheads of loading and unloading of vector registers, especially the saving and restoring of the relatively large contents of registers whenever a context switch takes place; for example, loading a single vector register on a Cray machine requires about 20 cycles. Indeed, optimizing the use of vector registers is arguably the most important problem in using such machines [1]. Some vector machines, such as those in the Fujitsu VP and Amdahl series, try to get around the first difficulty by allowing vector-register length to vary, but this is only within a small range and is achieved by reconfiguring the vector-register file so that the number of registers available decreases as the register length increases [25, 28].

Another important issue in the design of vector machines is how hazards are detected and resolved. The basic problem here is similar to that for scalar computations (Chapter 5), but now consideration must also be given to the level at which the synchronization of producers and consumers of data is done: it can be on whole vectors or on individual elements of vectors. If the synchronization is on whole vectors, then the processor's idle-time is determined primarily by the lengths of the vectors involved; on the other hand, if synchronization is on individual elements, then the idle-time is determined primarily by algorithmic considerations. Fine-grain synchronization is therefore desirable, but it is more costly to provide, and most machines do so only in special cases: the Cray and ETA machines through *chaining*, in which vector results are passed on an element-by element basis, through registers, from one functional unit to another that uses them as inputs; and the CDC Cyber and ETA machines implement short-stopping (as described above) and programmer-controlled *linking* (forwarding), which is analogous to chaining. These types of arrangements are easiest with register-to-register machines, in which intermediate results are stored in named registers, and this explains why in the Cyber 205 the limited hardware technique of shortstopping must be supplemented with one that requires the intervention of the programmer. For either type of architecture, even the limited fine-grain synchronization has been shown to enhance performance: on both Crays and Cybers performance on linked and chained operations is twice the normal rate. (A description of a vector-machine in which all synchronization is at a fine granularity is given in [18].)

6.2 STORAGE AND ADDRESSING OF VECTORS

A vector may be distinguished as being either *dense* or *sparse*, according to whether or not it has a preponderance of useful values, i.e. values that cannot be excluded because of their numerical significance — for example, non-zeros in many applications.[1] Evidently computing with, or storing, all the elements of a sparse vector is not an efficient use of storage, memory bandwidth during access, and processor bandwidth during computation; and it should be remarked that there exist many important scientific applications that use very large sparse vectors. Sparse vectors therefore require special attention with regard to storage, whose use should be minimized, and access. Dense vectors will be stored in their entirety, but even here attention needs to be paid to access, as this can be irregular: for example, a conditional operation on a vector will generally involve only some particular elements — and in no particular order.

There are two main methods for the storage of sparse vectors, both of which involve storing the non-zero elements in a shorter dense vector, along with

---

[1]In what follows, we shall, for simplicity, assume that only zeros are not significant.

Table 6.2.   Cyber 205 vector instructions for sparse and random access

Instruction	Description
MASK A, B, C, Z	If $Z_i = 1$, then set C_i to A_i; else set C_i to B_i. Either operand may be broadcast. Operation length is determined by Z.
COMPRESS A, C, Z	C consists of elements of A (in order) that correspond to 1s in Z. Operation length is determined by Z.
COMPRESS A, B, C, Z (ARITHMETIC) may be	Compare corresponding elements of A and and B. Where $A_i \geq B_i$, copy A_i to C and store 1 in Z_i; otherwise store 0 in Z_i. B a broadcast vector. Operation length is determined by A.
MERGE A, B, C, Z	C consists of elements of A (in order) for 1s in Z and elements of B (in order) for 0s in Z. Operation length is determined by Z.
DECOMPRESS A, B, C, Z	Similar to MERGE, but skip one element of B for each element of A copied to C.
EXPAND A, B, C, Z	Similar to MERGE, but one of A or B is a broadcast vector.
COMPAREEQ A, B, Z	Compare corresponding elements of A and B. If $A_i = B_i$, store 1 in Z_i; else store 0 in Z_i. Either operand may be broadcast Operation length is determined by Z.
COMPARENE, COMPAREGE, COMPARELT	Similar to COMPAREEQ; tests are for $\neq$ $\geq$, and $<$.
ADD A, B, C, X, Y, Z	Add A_i (if X_i is 1) or zero (if X_i is 0) to B_i (if Y_i is 1) or to zero (if Y_i is 0).
SUB, MULT, DIV	Subtract, multiply, and divide; similar operation as for ADD.
GATHER A, B, C	Transfer elements of B into C, where B is indexed by elements of A. Length of C is equal to length of A.
SCATTER A, B, C	Transfer elements of B into C, where C is indexed by elements of A. Length of C is equal to length of A.
INTERVAL A, B, C	Result C contains numbers that start with value in A and successively increase by that in B.

broadcast vector = operand vector obtained by on-the-fly replication of named scalar
A, B = operand data vectors; C = result data vector; X, Y, Z = bit vectors

an additional vector that, in essence, allows an on-the-fly reconstruction of the original vector; that is, during movement for computation, the additional vector is used to determine the positioning (relative to the original vector) of the elements. The additional vector may be a *bit vector* (of 0s and 1s) that is as long as the original data vector and in which a 0 indicates a zero element in the original vector and a 1 indicates a non-zero element, or it may be an *index vector* that contains the indices (in the original data vector) of the non-zero elements and whose length is therefore determined by the number of are non-zero elements. The choice of which of the two to use may depend on the particular programming circumstances and is discussed Section 6.4.

The process of creating a shorter, dense data vector together with a bit vector from a sparse vector is commonly referred to as *compression*, and the obverse is known as *expansion*. A typical vector machine will have these instructions in its instruction-set, along with a number of other instructions on sparse vectors. Table 6.2 lists the sparse-vector instructions of the CDC Cyber 205, which has two operations for compression and three for expansion; and Figure 6.5 demonstrates the application of some of these. In typical use, the boolean operations are first used to generate bit vectors that are then used in subsequent operations. Thus, for example, a compression operation in the Cyber 205 would consist of a COMPARE followed by a COMPRESS; and an efficient arithmetic operation might be realized on sparse vectors by compressing the operands, performing the operation on the resulting shorter vectors, and then expanding (or not) the intermediate result.

The Cray computers also implement a variety of compression and expansion operations, although these are register-based and therefore more constrained than those in the Cyber 205. Since processor vector operations are carried out only on vector registers (each of which holds 64 elements), with results returned to registers, a 64-bit Vector Mask (VM) register, in which each bit corresponds to a vector-register element, is sufficient for controlling sparse operations. An example of a compression on a Cray is given in Figure 6.6, and Table 6.3 lists all of the related instructions in that architecture [7].

Although index vectors can be also be used for the storage of sparse vectors, they are more commonly used to realize random access to substructures of dense vectors. The operations here are known as *gather* (for Reads) and *scatter* (for Writes). Gather is a realization of $B[i] := A[I[i]]$, and Scatter is a realization of $B[I[i]] := A[i]$, where A and B are operand vectors and I is an index vector; examples are shown in Figure 6.7. In a memory-to-memory machine, such as the CDC Cyber 205, all three vectors (A, B, and I) would be in memory, whereas in a register-to-register machine, such as a Cray, for Scatter, A and I would be held in vector registers, and B and I would be similarly stored for Gather.

Given that the length of the operation in Gather and Scatter is determined by the length of I, which may be shorter or longer than A or B, it is evident that two operations may be used for compression and expansion. However, we saw in Chapter 3 that the best access rates for vectors in interleaved memory is achieved when the stride is constant; so we should expect that the variable stride in Gather and Scatter will result in relatively poor performance, and indeed this is so: In the CDC Cyber 205 the best performance on Gather and Scatter is less than one-third of that for almost all other operations; and on

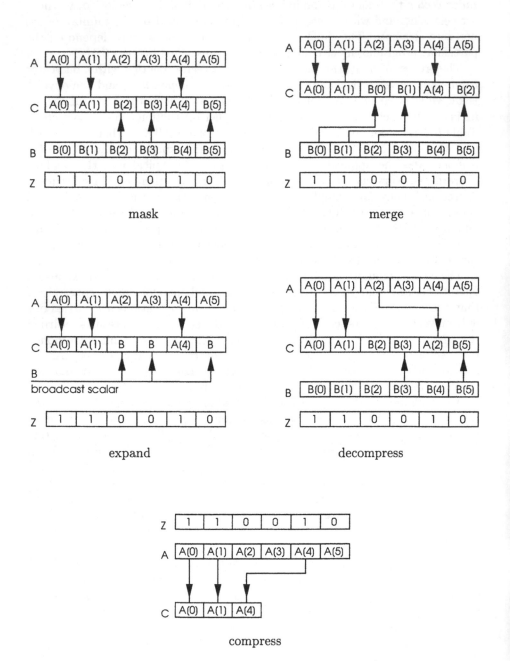

Figure 6.5.   Examples of Cyber 205 sparse-vector operations

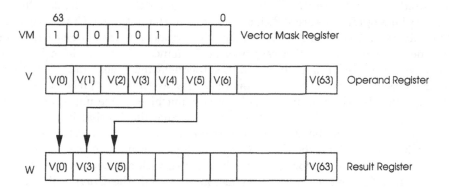

Figure 6.6.   Example of compression on Cray computer

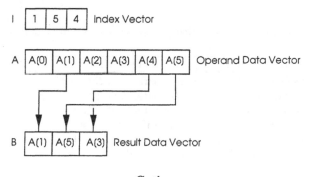

Gather

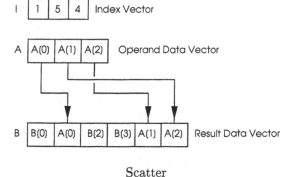

Scatter

Figure 6.7.   Gather and Scatter operations

that machine, whereas performance on 32-bit operands is double that for 64-bit operands for most operations, there is no change for these two [21, 23]. Nevertheless both operations are sufficiently important that although Cray computers initially did not have built-in vector support for them — they were carried out by the scalar processor, with very poor performance — this was later found to be a serious omission and corrected in subsequent machines. Indeed, the two operations are sufficiently important that they are now routinely included in performance-benchmark programs for vector machines. One important difference between the CDC machines and the Cray machines is that in the former, best performance can be attained only with unit-stride vectors — to access vectors with other strides, the slower Gather and Scatter must be used — whereas with the latter, best performance can be reached with any constant stride. We should therefore expect that, other factors excluded, the Crays will perform better over a wide range of operands and operand-access patterns. In fact, even Gather and Scatter perform reasonably well on recent Crays — at about one-half the rate for most other operations [34].

Table 6.3.   Cray instructions for sparse vectors and random-access

Instruction	Description
SETMASK S	Move contents of S to VM.
CLEARMASK	Clear contents of VM.
MOVEMASK S	Move contents of VM to S.
SETMASKZ V	Set bit i if VM if $V_i = 0$, else clear.
SETMASKNZ,	Similar to SETMASKZ, but test for
SETMASKPOS,	$\neq 0$, ≥ 0, and < 0.
SETMASKNEG	
SETMASKZV U, V	If $V_i = 0$, set bit i of VM and copy contents of V_i to U_i.
SETMASKNZV,	Similar to SETMASKZV, but test for
SETMASKPOSV,	$\neq 0$, ≥ 0, and < 0.
SETMASKNEGV	
MOVES S, V, W	If bit i of $VM = 1$, copy contents of S to W_i; else copy V_i to W_i.
MOVEV U, V, W	If bit i of $VM= 1$, copy U_i to W_i; else copy V_i to W_i.
MERGE V, W	If bit i of $VM= 1$, copy V_i to W_i, else store 0 in W_i.
GATHER A, U, V	Add contents of A to U_i to form address. Load from the memory address into V_j. $i, j = 0, 1, 2, \ldots$
SCATTER A, U, V	Add contents of A to U_i to form address. Store V_j to memory address. $i, j = 0, 1, 2, \ldots$

U, V, W = vector registers; A = an address register

VM = vector mask register; S = scalar register

6.3   INSTRUCTION SETS AND FORMATS

Tables 6.2 and 6.4 together list the vector instructions of the CDC Cyber 205 and its descendant, the ETA 10. The vector instructions have the general format shown in Figure 6.8; within this general format, there are variations, according to the class of instruction, and not all of the various field are used for all instructions. Each of the six operand-specification fields consists of eight bits that name one of 256 scalar registers that contains the information indicated in the figure. For some instructions there is also an additional operand specification that is implicit: if register c is named in the C field, then register $c + 1$ is also an operand. The detailed information enoded in the various fields are as follows. The F field specifies the main operation code. The G field specifies additional information, such as the operand width, whether one or both (and which) of the operands is a broadcast vector, whether the operands are signed or unsigned, and whether a named bit vector operates on 0s or 1s. And the information contained in a named source-operand scalar register is either a scalar operand (if the vector is broadcast) or is interpreted according to Figure 6.9.

Table 6.4.   Vector instruction set of CDC Cyber 205

Instruction	Description
ADD A, B, C	Add A and B, with result to C. Either operand may be broadcast.
SUB, MULT, DIV	Subtract, multiply, divide. Similar to ADD.
TRANSMIT A, C	Copy A to C. A may be broadcast.
SHIFT A, B, C	Shift A_i by B_i; result into C_i. Either operand may be broadcast.
ABSOLUTE A, C	Set C_i to $|A_i|$. A may be broadcast.
FLOOR A, C	Set C_i to $\lfloor A_i \rfloor$. A may be broadcast.
CEILING A, C	Set C_i to $\lceil A_i \rceil$. A may be broadcast.
EXPONENT A, C	Exponent of floating-point A_i to C_i.
TRUNCATE A, C	Set C_i to integer of largest magnitude $\leq A_i$.
PACK A, B, C	Set C_i to floating-point number with exponent A_i and significand B_i.
LOGICALOP A, B, C	Set C_i to A_i **op** B_i, where **op** is one of AND, OR, XOR, STROKE, PIERCE IMPLY, INHIBIT, EQUIVALENCE.
EXTEND A, C	Extend 32-bit floating-point A_i to 64-bit floating-point C_i.
CONTRACT A, C	Contract 64-bit floating-point A_i to 32-bit floating-point C_i.

broadcast vector = operand vector obtained by on-the-fly replication of named scalar

A, B = operand vectors; C = result vector

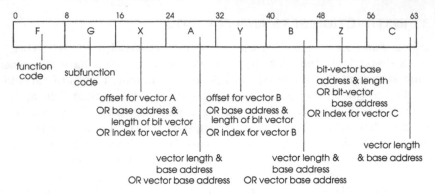

Figure 6.8.   Cyber 205/ETA 10 vector-instruction format

In a Cray processor, all functional units take operands from registers and return results to registers; memory can be accessed by only a few types of instruction that load into or store from the operating registers. The instructions and their formats are therefore much simpler and smaller than those in the CDC machines. The registers accessible to the functional units are eight 32-bit Address registers, eight 64-bit Scalar registers, and eight 64-element Vector registers (with each element a 64-bit one). Two other registers are used in vector computations: a 64-bit Vector Mask register and a 7-bit Vector Length register. Tables 6.3 and 6.5 together list the vector (and related) instructions of the Cray X-MP [7]. The primary formats used for these instructions are shown in Figure 6.10; some fields are unused in some instructions.

6.4   PROGRAMMING TECHNIQUES

In this section we shall briefly look at *vectorization techniques* that may be used by compiler or programmer to facilitate the generation of code for a vector-pipelined machine [13, 39]. We examine two techniques in some detail and a

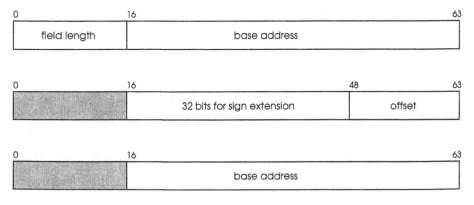

Figure 6.9.   Cyber 205/ETA 10 vector-address format

Table 6.5.   Vector instruction set of Cray X-MP

Instruction	Description
CLEAR V	Clear elements of V.
CLEAR A, V	Clear element (A) of V.
MOVELA VL, A	Move (VL) to A.
MOVEVS V, A, S	Move element (A) of V to S.
MOVEVV U, V	Move elements of U to V .
MOVEAL A, VL	Move (A) to (VL).
MOVE1L 1, VL	Move 1 to (VL).
STOREC A, C, V	Store V to consecutive addresses, starting at (A), constant increment (C).
STORE A, V	Store V to consecutive addresses, starting at (A).
LOADC A, C, V	Load V from consecutive addresses, starting at (A), constant increment (C).
LOAD A, V	Load V from consecutive addresses, starting at (A).
ADDSV S, U, V	Add (S) to (U_i), result into V_i.
DIFFSV, MULTSV	Difference, multiplication, logical, reciprocal.
LOGICSV, RECIPSV	Similar to ADDSV.
ADDVV U, V, W	Add (U) to (V_i), result into W_i.
DIFFVV, MULTVV	Difference, multiplication, logical, reciprocal.
LOGICVV, RECIPVV	Similar to ADDVV.
SHIFTL A, U, V	Shift (U_i) left by (A) places, result to V_i.
SHIFTR A, U, V	Shift right.
POPUL U, V	Population count of U, result to V.

U, V, W = vector registers; A, C = address registers
S = scalar register; (*) = contents of *

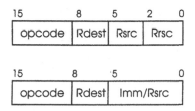

Figure 6.10.   Cray vector-instruction formats

few others more briefly; practical vectorization consists of the application of a number such techniques in combination.

As indicated above, conditional expressions result in irregular access patterns and therefore require the use of Compress/Decompress or Scatter/Gather operations if they are to be vectorized. There are two main methods for doing so: the *masking* (or *bit vector*) method and the *short-vector* method. In the former, bit vectors (masks) are used to realize selective operation on the operands; and in the latter, index vectors are used to produce short vectors of the operand-elements to be operated upon. For examples, of the application of these methods, we give the sequences of instructions that would be generated in the Cyber 205 for the code fragment

$$\textbf{for } i:=1 \textbf{ to n do}$$
$$\textbf{if } A[i] \neq 0 \textbf{ do}$$
$$B[i]:=C[i]+D[i]$$

Table 6.6.   Methods for processing conditionals

Code Sequence	Vector Length
COMPARE A, 0, Z	n
ADD C, D, B, Z, Z	n

(a) masking (bit vector) method

Code Sequence	Vector Length
COMPARE A, 0, Z	n
COMPRESS C, U, Z	n
COMPRESS D, V, Z	n
ADD U, V, W	$n * d$
DECOMPRESS W, B, B, Z	n

$d = $ data density, i.e. ratio of non-0s to 0s

(b) short-vector method w. Compress/Decompress

Code Sequence	Vector Length
COMPARE A, 0, Z	n
INTERVAL 1,1, J	n
COMPRESS J, I, Z	n
GATHER I, C, U	$n * d$
GATHER I, D, V	$n * d$
ADD U, V, W	$n * d$
SCATTER I, W, B	$n * d$

(c) short-vector method w. Gather/Scatter

The use of the masking method is shown in Table 6.6(a): the first instruction generates a bit vector, containing 1s where the condition is met and 0s elsewhere, and the second instruction adds operands elements for those positions in which the bit vector has 1s. The short vector method may be realized in the two ways, as shown in Table 6.6(b) and Table 6.6(c). In (b) the data to be operated upon are compressed into short vectors, the arithmetic operation is carried out, and the intermediate result then expanded into the result vector; (c) is largely similar, but with the generation of an index vector, gathering in place of compression, and scattering in place of expansion. The important difference between the two short-vector methods is that in the first each COMPRESS is on a full length vector, whereas in the second, after the initial compression of the index vector, each GATHER is on a short vector.

The choice of which to use between the masking method and the short-vector method depends on a number of factors, for each particular case: the machine architecture, the data density in the operands, and the number of arithmetic operations to be carried out on those operands, and so forth. All operations in the masking method are always on full-length vectors, whereas with the short-vector method the number of elements on which arithmetic is performed depends on the data density (i.e. on the length of the short vectors). Thus low density favours the short-vector methods, provided the density and number of arithmetic operations to be carried out is such that the overheads of compression can be amortized. Between the two short-vector methods, the choice depends on the relative costs of Compress/Decompress and Gather/Scatter operations. On a memory-to-memory machine the memory stride in Gather/Scatter operations is irregular, and the two operations will therefore typically run much slower that Compress/Decompress; this suggests that Compress/Decompress will be the better method, unless the vectors involved are very sparse.

Another particularly useful vectorization technique is *loop interchange*, which involves changing the order of the indexing statements in nested loops. The three main uses of this technique are to facilitate vectorization where none would otherwise be possible, to improve execution time by producing longer vectors (and therefore amortizing start-up time), and to change access stride. For example, the code fragment

$$\textbf{for } j:=1 \textbf{ to } n \textbf{ do}$$
$$\textbf{for } i:=1 \textbf{ to } n \textbf{ do}$$
$$A[i+1, j]:=A[i, j]*B[i]+C[i, j]$$

cannot be vectorized as it stands, because of the data dependences (read-after-write) between iterations of the inner loop: for example, for the first iteration of the outer loop we have

```
A[2,1]  := A[1,1]*B[1]+C[1,1]
A[3,1]  := A[2,1]*B[2]+C[2,1]
A[4,1]  := A[3,1]*B[3]+C[3,1]
```

in which every operation depends on the result of the preceding operation. But if we interchange the two indexing statements that control the iterations, then such hazards are eliminated, as the iterations of the inner loop are now independent of one another:

```
A[2,1] := A[1,1]*B[1]+C[1,1]
A[2,2] := A[1,2]*B[2]+C[1,2]
A[2,3] := A[1,3]*B[3]+C[1,3]
```

and the code can now be vectorized by replacing the loop body with a single vector instruction. On the other hand, the code fragment

```
for j:=1 to 1000 do
    for i:=1 to 10 do
        A[i, j]:=A[i, j]*B[i]+C[i, j]
```

is already vectorizable; however, loop-interchange improves upon it by allowing a change from many operations on short vectors to a few operations on long vectors. And in this piece of code

```
for l:=1 to n do
    for j:=1 to m do
        A[i, j]:=A[i, j]*B[i]+C[i, j]
```

loop-interchange would produce an access stride of unit where there is none.

Other commonly used vectorization techniques are as follows. *Scalar promotion* involves converting a scalar operand into a vector one from code such as

```
for l:=1 to n do
begin
    S:=A[i]+B[i];
    if t≠0 then
        C[i]:=C[i]*S
end
```

Induction-variable recognition involves detecting induction variables that are used as vector subscripts and therefore facilitates the determination of the vector-access patterns; for example, j is an induction-variable in the code fragment

```
j:=n;
for l:=1 to n do
begin
    A[j]:=B[i];
    j:=j-1
end
```

In *preloading*, vectorization is facilitated by scheduling LOADs and STOREs in a way that resolves potential hazards — for example, loading A[i+1] before storing A[i] in the code for the fragment

```
for 1:=1 to n do
begin
        A[i]:=B[i];
        B[i]:=A[i]+A[i+1]
end
```

Lastly, in *loop collapsing* longer vectors are produced by eliminating the nesting of loops, e.g. viewing a structure as a one-dimensional 1×10000 vector instead of 100×100 matrix. Further discussion of vectorization techniques will be found in [39], on which this section is based, and in [13].

6.5 PERFORMANCE

Re-arranging the terms in Equations 6.1 and 6.2, we see that the time for a vector operation consists of some overheads for initialization plus the time to actually carry out the operation on the elements of the operands. So, if we let t_{st} denote the overhead time and t_{elt} denote the time required to process one set of operand elements, then we may replace Equations 6.1 and 6.2 with the generic one

$$T_V \ = \ t_{st} + nt_{elt} \tag{6.3}$$

Neither t_{st} nor t_{elt} is constant — even for a given machine. (Table 6.7 gives some measured values [4].) In practice the need to re-position data as it goes into the pipelines means that both times will vary according to the operation at hand; for example, operations on sparse vectors will typically take longer than corresponding operations on on full vectors. In the best case, t_{elt} will be equal to the pipeline cycle time, t_p, and so one measure of performance is the *maximum throughput*, r_∞, which in the simplest case is the inverse of cycle time (Chapter 1). In the case where several pipelines units can be in operation concurrently, t_{elt} is (ideally) reduced proportionately, and a corresponding multiplicative factor is introduced into r_∞. Thus, for example, the Cray-1 has a cycle time of 12.5ns, which yields a normal peak performance of 80 MFLOPS per functional unit, and where chaining is possible two such units can be usefully operated in parallel; so in the best case $t_{elt} = 12.5$ns and $r_\infty = 80$ MFLOPS normally, and with chaining $t_{elt} = 6.25$ns and $r_\infty = 160$ MFLOPS. Unless otherwise specified, we shall in this section assume the simple case of just one arithmetic pipeline.

The peak performance is, however, not a very useful measure: it assumes that the only orders processed are vector ones, and it does not take into account such factors as start-up time and vector length. Moreover, r_∞ is a technologically dependent parameter that gives little information on the *quality* of an architecture. A more useful measure of performance is the *half-performance vector length*, denoted by $n_{1/2}$ [16]. This is defined as the vector length needed to reach $r_\infty/2$. From Equation 6.3, the average performance on vectors of length n is

$$\frac{n}{t_{st} + nt_{elt}}$$

Table 6.7.   Start-up and per-element times (cycles)

| Vector | Cyber 205 | | Cray-1 | | |
Operation	t_{start}	t_{elt}	t'_{start}	t_{strip}	t_{elt}
V:=V+S	102	1	72	22	2
V:=V+V	102	1	72	27	3
V:=V*S	98	1	72	24	2
V:=V*V	98	1	72	28	3
V:=V+S*V	158	1	72	27	4
V:=V+V*V	158	1	72	36	3

2-pipeline Cyber, S=scalar, V=vector

and when half the peak performance is attained, we have

$$\frac{r_\infty}{2} = \frac{n_{\frac{1}{2}}}{t_{st} + n_{\frac{1}{2}} t_{elt}}$$

which, assuming a best case of t_{elt} (i.e. the clock cycle time) yields, from $t_{elt} = 1/r_\infty$

$$n_{\frac{1}{2}} = r_\infty t_{st} \qquad (6.4)$$

$$T_V = \frac{n_{\frac{1}{2}} + n}{r_\infty}$$

Equation 6.4 gives another interpretation of $n_{1/2}$: it is a measure of how much useful work is lost due to the start-up time. Moreover, $n_{1/2}$ is a technology-independent parameter. To see this, suppose the cycle time is reduced by a factor of λ and that this produces a new half-performance vector length, $\bar{n}_{1/2}$, a new peak performance, $\bar{r}_\infty$, and a new start-up time, $\bar{t}_{st}$. Since all times are reduced by the same factor, we have

$$\bar{n}_{\frac{1}{2}} = \bar{r}_\infty \bar{t}_{st}$$

$$= \frac{\lambda}{\tau_p} \frac{t_{st}}{\lambda}$$

$$= r_\infty t_{st}$$

$$= n_{\frac{1}{2}}$$

The parameter $n_{1/2}$ is useful for evaluating the quality of an architecture and for comparing different architectures, but it is not sufficient to answer other important questions such as the following: For a given machine, at what point is using the vector processor more beneficial than using just the scalar processor? For a given application, how does the non-vectorizable part of the code (and, by implication, the compilation and specific operations encoded) affect the performance? Furthermore, the parameter on its own is not always useful, even in comparing machines of the same architecture but different implementations: for example, the Cray X-MP has a larger value of $n_{1/2}$ than the Cray-1 does but has smaller values of t_{st} and t_{elt}. We next consider how the two questions just posed can be answered.

The first question above can be answered by considering the performance parameter known as the *break-even vector length*, n_b, which is defined to be the vector length for which the time taken on the vector processor is equal to that taken on the scalar processor and beyond which the performance of the vector processor is better. Thus if the peak scalar processing rate is $r_{\infty s}$ and the peak vector processing rate is $r_{\infty v}$, then processing a vector of length n_b takes a time of $n_b/r_{\infty s}$ on the scalar processor. And from Equation 6.4 we have

$$\frac{n_b}{r_{\infty s}} = \frac{n_{\frac{1}{2}} + n_b}{r_{\infty v}}$$

which yields

$$n_b = \frac{n_{\frac{1}{2}}}{r_{\infty v}/r_{\infty s} - 1} \tag{6.5}$$

Since we want n_b to be as small as possible, Equation 6.5 implies that either $n_{1/2}$ should be made small or that the vector processor should be much faster than the scalar processor. (We shall see, however, that the latter is not necessarily desirable.) Values of the performance parameters discussed so far are shown in Table 6.8 for some machines.

Table 6.8.   Performance parameters of various machines (single pipeline)

Machine	$n_{1/2}$	$r_{\infty v}/r_{\infty s}$	n_b
Cray-1	10–20	13	1.5–3
Cyber 205	100	10	11
Star 100	150	12	12

To characterize the effect of the non-vectorizable part of some code (i.e. to answer the second question above), let us suppose that the average rate of scalar processing is r_s, so a single operation on the scalar processor takes an average time of $t_s = 1/r_s$, and that the average rate of vector processing is r_v, so a single operation on the vector processor takes an average time of $t_v = 1/r_v$. So, if the proportion of vectorizable operations is v (and the non-vectorizable proportion is therefore $1 - v$), then the average time per operation is given by

$$t = vt_v + (1 - v)t_s$$

If we define the *average speed-up*, S, of using both vector and scalar processors, instead of just the scalar processor, to be the ratio t_s/t, then we have

$$S = \frac{t_s}{vt_v + (1-v)t_s}$$

$$= \frac{1}{vt_v/t_s + (1-v)}$$

$$= \frac{1}{vr_s/r_v + (1-v)}$$

$$= \frac{1}{1 + v(1 - r_s/r_v)}$$

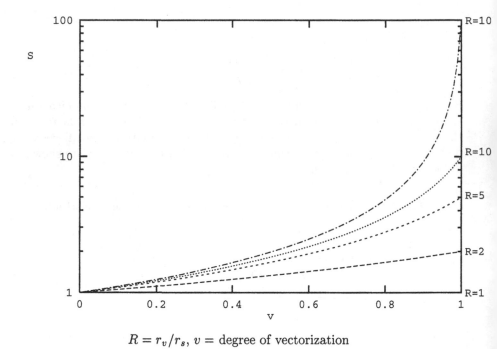

$$R = r_v/r_s, \ v = \text{degree of vectorization}$$

Figure 6.11.   Speedup of a vector machine

Figure 6.11 shows a plot of the speed-up for various values of r_v/r_s and v. As we should expect, for a given value of r_v/r_s, the speed-up improves with the degree of vectorization; and for a given level of vectorization, the speed-up increases with r_v/r_s. What is more interesting, however, is the fact that for a given level of vectorization, as r_v/r_s increases, the rate at which S increases is not proportionate and gets smaller. So a change in v is a more effective way of improving performance than a change in r_v/r_s. This is what we should expect:

the effect of the magnitude of v is clearly evident if we assume an infinitely fast vector processor, in which case $S = 1/(1-v)$; that is, in the limit, as the vector processor gets faster, the overall performance is determined by the performance over the non-vectorizable code.

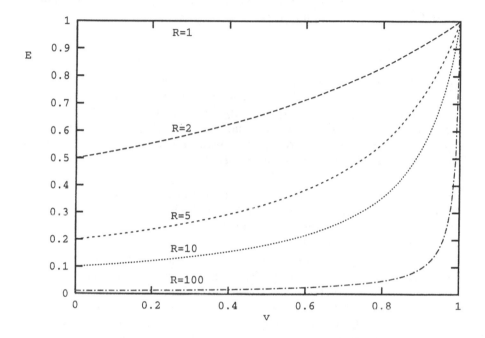

Figure 6.12.   Relative speed-up of a vector machine

The fact that ultimately performance is limited by part of the code that is executed at the slowest rate, and that beyond some point it is not worthwhile to improve the performance of the vector processor without also improving the performance of the scalar processor, is known as *Amdahl's Law*. (The law is also applicable to more general parallel-processing systems, in which v corresponds to the proportion of parallelizable code.) Another way to see the effect of Amdahl's Law is look at the *relative speed-up* (or *efficiency*), S_{rel}, of the system; this is defined as the ratio of the average speed-up to the maximum possible speed-up, i.e. $S_{rel} = S/(r_v/r_s)$. Ideally, we want S_{rel} to be close to unity. But, as Figure 6.12 shows, as r_v/r_s increases, either S_{rel} drops sharply or v must be increased sharply just to retain the same efficiency; the sharp rise as v approaches unity is known as *Amdahl's Wall*.

6.6 CASE STUDIES

This section consists of two case studies: the CDC Cyber 205, as an example of a memory-to-memory machine, and the Cray X-MP, as an example of a register-to-register machine [7, 8, 23, 37]. An alternative approach to implementing a memory-to-memory pipeline will be found in the secondary pipeline of the Manchester MU5 [27]; and other examples of register-to-register machines are those from NEC, Fujitsu, and Hitachi [28, 30, 38].

6.6.1 CDC Cyber 205

The Cyber 205 is one implementation of the Cyber 200 memory-to-memory vector architecture [8, 23]. A basic machine consists a very large and highly-interleaved memory (Chapter 3), up to eight input/output ports (with an option for eight more), a scalar processor, and a vector processor with one pipeline (and an option for one or three more pipelines). The basic units for data are words and 512-bit *super words* (or *swords*).

Scalar processor   The scalar processor has the organization shown in Figure 6.13. Its main functions are the execution of all scalar instructions and instructions that carry out various system functions. The execution is carried out in five independent functional units.

Instructions enter the processor through the Memory Control Unit and the Instruction Stack, a lookahead buffer of sixty-four words that at any moment holds a combination of 32-bit and 64-bit instructions, with a maximum number of 128 32-bit ones or 64 64-bit ones. The stack is essentially a larger version of that used in the CDC 7600 (Chapter 4): it consists of eight swords than are addressed independently, thus allowing disjoint code segments to be stored and for out-of-stack branches to be processed without flusing the stack. The Memory Control Unit has three main parts: the Priority Unit (see Section 3.1.3) is the interface to the memory, the Read-Next-Sword (RNS)/Branch Unit controls the loading of the Instruction stack and also executes branch instructions, and the Associative Unit translates virtual addresses into real ones. The Priority Unit takes memory-access requests from the various units in the processor, filters out the invalid ones, and processes the rest, with priority arbitration in the event of simultaneous requests. In the event that an addressed memory bank is busy, the request is repeated, either by the Priority Unit or by the original source. Requests are of two types: *immediate-issue* and *delayed-issue*. The former type consists of all READ requests (except I/O READs) and WRITE requests of one word or half a word. The latter type consists of single-sword or double-sword WRITEs that are issued from the Priority Unit four cycles later than an immediate-issue request; the extra time is for the assembly into swords before accessing memory. The RNS/Branch unit consists of two main parts: the RNS part controls the loading of the Instruction Stack and maintains lookahead, by always fetching two swords beyond the sword containing the instruction currently being decoded; and the Branch part carries out the branch-condition testing and the transfer of control. The Associative Unit translates addresses by using a table whose contents map the entire virtual memory[2] onto real memory: each entry of the table corresponds to one page frame and holds the address of the virtual page that is currently mapped into that frame. In principle, the table contains an entry for every virtual page, is nominally associatively addressed, and is managed using a true least-recently-used algorithm, but cost and performance reasons dictate that it actually be implemented otherwise: Only the top sixteen locations (for the most recently used pages) are held in associative registers. The rest are held, in order of use, in memory and searched, when necessary, using special hardwired instructions; the principles of locality ensure that this arrangement performs adequately in practice.

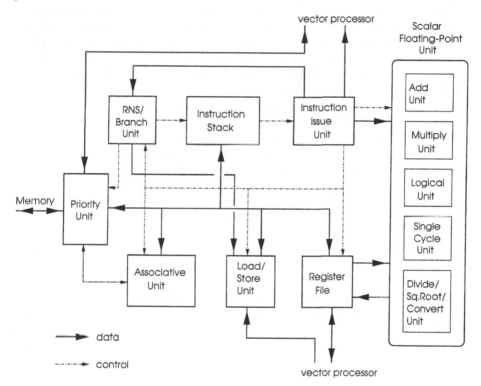

Figure 6.13. Cyber 205 scalar processor

The Instruction Issue Unit takes non-branch instructions from the Instruction Stack, decodes them, and then issues them to the scalar functional units or to the vector processor, as appropriate. The parallel operation of the scalar and

---

[2]The CDC/ETA machines are the only supercomputers with virtual memory; others avoid them on performance grounds. Also note the unusual and variable page sizes in the Cyber: 4K-bytes, 16K-bytes, 64K-bytes, and 512K-bytes.

vector processors is as follows. No instruction is issued if doing so will result in register-write conflicts or in the scalar processor generating memory references while the vector processor is busy. When the Instruction Issue Unit decodes a vector instruction, and the vector processor can accept the instruction, the function code and the contents of all registers specified in the instruction are forwarded to the processor, and all registers than can be modified during the execution of the instruction (e.g. registers used for indexing) are reserved. If the vector processor is not ready, then the Instruction Issue Unit is stalled until released by the vector processor when it does become free.

Hazards are detected and resolved by the Instruction Issue Unit as follows. RAW and WAW hazards are detected by holding the destination-register addresses of issued (but uncompleted) instructions in sixteen Result Address Registers and then comparing the contents of these with the operand-register addresses in the next instruction due to be issued; a hazard is signalled if there is a match. Register-write conflicts (the structural hazards) are detected by checking each destination-register address against a timing chain of result positions. RAW and WAW hazards are resolved by stalling the pipeline; and WAR hazards do not occur, as an instruction is dispatched/issued only when its operands are available. The instruction pipeline is also stalled if the results of two instructions, issued at different times and with different execution times, arrive at the register file at the same time. The register file itself consists of 256 64-bit registers that are used for addressing, indexing, to hold vector-operand lengths, and as source and destination registers for scalar instructions; the file can process two READs and one WRITE in each cycle. In implementation, the register file is replicated in order to reduce access-conflicts, and it also includes a bypass path that enables a scalar result produced by one instruction to be made available as an operand to a subsequent instruction at the same time as the result is written into a register, thus eliminating the delay of a Read operation.

The Load/Store unit, under the control of the Instruction Issue Unit, accepts addresses and transfers data between registers and main memory; to do this, the unit uses six registers to buffer requests and then processes them in correct order. The unit is capable of streaming requests at a rate of one LOAD request per cycle and one STORE request in every two cycles, assuming no register-write conflicts. A stream of n LOADs is processed in in $n + 14$ cycles and n STOREs in $2n + 13$ cycles.

Vector processor   The vector processor has seven main components (Figure 6.14): (i) the Vector Setup Unit, which controls the setup and rundown of all non-scalar instructions and also which executes certain control instructions; (ii) the Stream Addressing Pipeline, whose main function is to maintain a maximum flow of data-addresses from the processor to the memory; (iii) the Vector Stream Input Unit, which buffers and aligns input data; (iv) the Vector Stream Output unit, which buffers and aligns output data to memory; (v) the Vector String Unit, which carries out operations that involve string vectors (i.e.

vectors whose elements are of no more than sixteen bits each, aligned on bit-boundaries) to memory; (vi) up to four Vector Floating-Point Pipelines that carry out arithmetic and logical operations on vectors of word or half-word elements; and (vii) an Input/Output-and-Maintenance Unit (not shown).

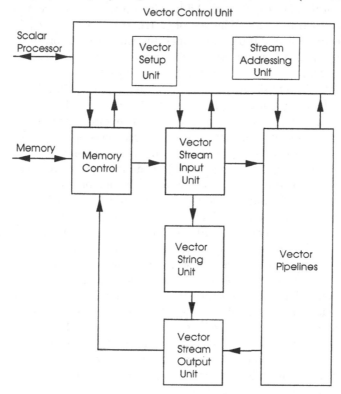

Figure 6.14.   Cyber 205 vector processor

The Vector Setup Unit consists of two main sections: the Setup section, which calculates the starting addresses and lengths of vectors, and the Execute unit, which sends out control signals to the rest of the vector processor. The hardware that makes up the two parts is controlled by microcode in the three memories: one for address and field-length calculations, one for execution monitoring, and one for decoded function bits. The Vector Setup Unit also contains nine counters that are used to restart a vector operation after an interrupt; the registers are updated during instruction execution and contain current field lengths, addresses, and extension data.

The scalar processor, after checking for hazards, issues instructions to the Vector Setup Unit without waiting for any instructions in progress to be completed but then waits for a "release" signal before continuing; the "release" signal is usually sent out as soon as the vector operation reaches an interruptible stage, but in a few cases it may be held up until the operation is completed. When a vector instruction is decoded in the Instruction Issue Unit, its operation code is sent to the Vector Setup unit, along with the contents of any required registers (containing vector addresses and lengths) in the register file. Next, the Execute subunit calculates and sends out the necessary control information, along with the starting addresses, basic length, and any necessary vector-extensions, to the Stream Addressing unit, and the overall length to the Vector Stream Input unit. The vector instruction is then ready for execution.

The Vector Stream Input unit operates (in conjunction with the Stream Addressing Pipeline) as follows. For each vector to be read, a buffer to hold the data is allocated in the Vector Stream Input unit, and a corresponding counter (initialized to the number of 128-bit words to be read) is allocated in the Stream Addressing Pipeline. The Stream Addressing Pipeline then generates sword or double-sword addresses and makes memory requests (accompanied by additional information, such as starting quarter-word or half-word) via the Priority and Associative units. (The addresses are generated at a maximum rate of one every four cycles for swords or double-swords and one every cycle for words and half-words; the rate is lower if memory-bank conflicts occur.) Data returns into the buffer, and the contents of the words-fetched-counter are decremented as the data is removed for use. The delivery of the data may be suspended if the allocated buffer is full, but the buffer has the capacity to hold all data that has been requested but not used. The Stream Addressing Pipeline is designed to receive and start each operation independently of other operations. This is accomplished by assigning priorities to requests — e.g. input operands have higher priority than output operands — and requiring that no request be processed if a higher-priority request to the same bank has been initiated within the preceding three processor cycles or will be satisfied within the next three cycles. (A bank requires four cycles to satisfy a given request.) Thus each request goes through a seven-cycle timing chain (three pre-initiation cycles, one initiation cycle, and three post-initiation cycles) before being forwarded to main memory.

The Vector Stream Output unit similarly buffers data being written into memory. This unit has two main sections: one for output from the arithmetic pipelines and register file and one for the output from the String unit. The pipeline-and-register section accepts word and half-word data (at a rate of up to 128 bits per cycle for a one-pipeline or two-pipeline configuration and 256 bits per cycle for a four-pipeline configuration) from the functional units and aligns them on word boundaries for input to memory; the other unit accepts string data and does the same thing on half-word boundaries. The procedure in the pipeline-and-register section is carried out in five subunits: (i) Output Selection, in which the next source of data is selected from the register file and (up to) four arithmetic pipelines;(ii) Compression, in which unwanted data are removed in sparse-vector operations; (iii) Alignment, in which a 256-bit bus is used to align a half-word of data (into one of eight possible positions within a half-sword) for writing into memory; (iv) Buffering, in which data are assembled into sword or double-sword units before being sent to memory; (v) and error detection and correction. The procedure in the string section consists of four steps: adjusting the string so that it starts on a half-word boundary and filling it so that it ends on a similar boundary; Data Selection, in which the input unit is selected; Alignment, in which data is shifted to the correct bit address; Merge, in which the string data is assembled into half-words; Buffering, in which the half-words are assembled into swords; and error detection and correction.

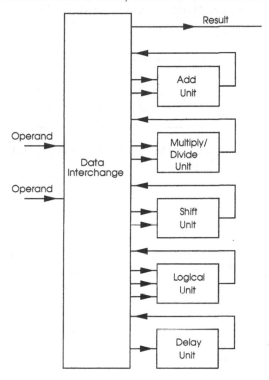

Figure 6.15.   Cyber 205 floating-point pipeline

A Vector Floating-Point Pipeline has the organization shown in Figure 6.15. Except for division and square-root operations, in each cycle the pipeline can process one pair of 64-bit operands or two pairs of 32-bit operands, thus doubling performance in 32-bit mode; and if two or four pipelines are used, then on a single operation they operate on alternate elements of the operand vectors. The basic arithmetic pipeline consists of five main functional units: Add, Multiply/Divide, Shift, Logical, and Delay. The Add unit carries out floating-point addition and subtraction and is a typical pipelined adder, but with two data-forwarding paths (Figure 6.16). The Multiply/Divide unit has the organization shown in Figure 6.17. The Multiply section consists of an array of carry-save adders, a carry-propagate (merge) section to assimilate the partial carries and partial sums and to complement intermediate results from the Divide section, and a unit that shifts significands and adjusts exponents; and the Divide section consists of a unit that complements operands — all operands for division and square-root are made positive on input — a divider, and a unit that calculates normalization-shift distances. The Shift unit consists of a two-stage shifter (Figure 6.18). The Logical unit (Figure 6.19) carries out pack/unpack, boolean, and masking operations. And the Delay unit (Figure 6.20) realizes a delay function, of a variable number of cycles, on data to be written to memory; the delay is implemented by offsetting, by these required number of cycles, the Read/Write addresses of the buffer.

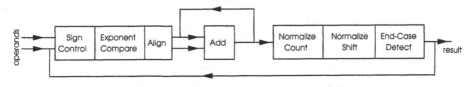

Figure 6.16.   Cyber 205 floating-point add unit

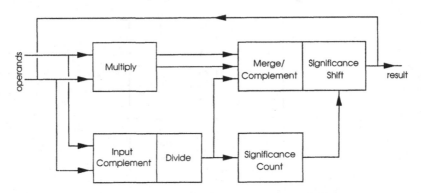

Figure 6.17.   Cyber 205 floating-point multiply/divide unit

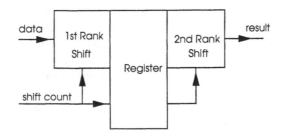

Figure 6.18.   Cyber 205 floating-point shift unit

The purpose of the last unit is to provide delay cycles for some operations that sequence data input; a linked-triad operation (i.e. one with three input operands and one output operand) is a typical example of such an operation.[3] In such an operation, evidently the third input must be delayed by the number of cycles taken to produce an intermediate output from the first two inputs.

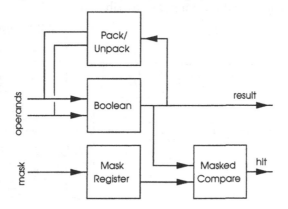

Figure 6.19.   Cyber 205 floating-point logical unit

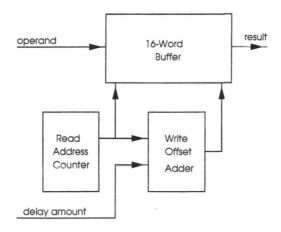

Figure 6.20.   Cyber 205 floating-point delay unit

6.6.2 Cray computers

The organization of a processor in the Cray X-MP is shown in Figure 6.21. The main components are a number of parallel functional units, a large and highly-

---

[3]The ETA machines have a special Shortstop Unit to facilitate automatic chaining. The Shortstop unit basically properly sequences results by buffering them. Cf. chaining in the Cray, below.

interleaved memory, and several sets of registers. Instruction-dispatch/issue is relatively simple: after decoding, an instruction proceeds only if none of the sources and destination it uses for operands is in use by another instruction; on dispatch/issue, an instruction places reservations on those registers that it will use and removes the reservations when it completes. The exception to this rule are in *tailgating* and chaining, as described below.

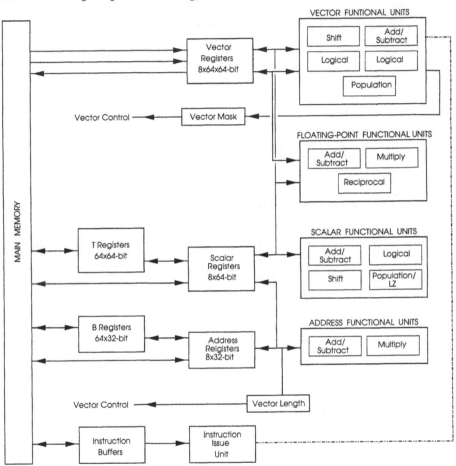

Figure 6.21.   Organization of a Cray X-MP processor

The main operating registers, i.e. those from which the functional units obtain their operands and to which they return their results, are eight 64-element×64-bit Vector registers, eight 64-bit Scalar registers, and eight 32-bit Address registers. The Vector registers are the main sources and destinations for vector operations; vector operations also use the Vector Mask register (for selective operation on vector elements, as described in Section 6.2) and the Vector Length register (which specifies how long a vector operation is). The Scalar registers mainly serve as sources and destinations for scalar and logical operations; however, they also provide operands for some vector operations. And the Address registers are used for memory addresses, indexing, shift counts, population counts, leading-zero counts, and so forth. Two additional sets of registers (the T and B registers) serve as buffers between main memory and the Scalar and Address registers. The T and B registers cannot be accessed directly by the functional units and are provided solely to allow rapid loading and unloading (via block-move operations) of the registers they support and to reduce the number of references between those registers and main memory. In effect, the T and B registers are a sort of programmable cache.

The Cray functional units are fully pipelined and, unlike those of the CDC machines, are mostly multifunctional. The units are all independent, but the number that can be usefully operated in parallel is constrained by the data-delivery paths and the particular sequence of operations to be carried out. The Address functional units carry out integer arithmetic on operands from the Address registers; neither unit checks for overflow, and multiplication returns only the least significant bits of a product. The Scalar functional units carry out integer arithmetic, logical operations, shifting, and leading-zero and population counting. Operands for these units come from the Scalar registers, except in shifting, in which the shift count is taken from an Address register; and except for leading-zero and population counting (for which the destination is an Address register), results are returned to the Scalar registers. The Floating-Point functional units are used for both scalar and vector floating-point operations: a scalar operation takes both operands from the Scalar registers and returns a result to the same; a vector operation takes one operand from a vector register, another operand from a vector or scalar register (unless the operation requires only one operand), and returns its result to a vector register. The Vector functional units carries out vector operations, of the type just described, and returns results to Vector registers. The functions of three of the functional units are self-explanatory; of the other two (Logical), one carries out operations that involve vector-masks, and the other carries out logical bit-by-bit operations. (Division in the machine is carried by by using a multiplicative normalization algorithm.)

The Cray machines all implement chaining. An example of chained sequence of operations (see Table 6.5) that involves several functional units would be the processing of the instruction-sequence

LOAD A0, v0
ADDVV v0, v1, v2
SHIFTL A1, v2, v3
LOGICVV v3, v4, v5

In effect, the chaining would build one long pipeline that starts at the memory, goes through the registers and functional units, and ends in register 5. Chaining may begin at any point in the production of the result data to be used as subsequent input; however, for full chaining to take place the consuming operation must be ready to use the first element of the chained vector by the time that is produced. The *tailgating* mechanism of the Cray 2 is a sort of obverse to chaining and allows the source vector register of an instruction that is near completion to be also used as a destination register by an instruction that is just starting, irrespective of the relationship between the reader and the writer of the register.

6.6.3 Japanese vector computers

The vector-supercomputer market is currently dominated by three Japanese manufacturers; NEC, Hitachi, and Fujitsu. The Japanese machines are all of the register-to-register type and therefore have many similarities with the Cray machines. In comparison with the Cray machines, their predominant features are more memory Load/Store units, larger vector instruction-sets, more vector registers (in some cases dynamically configurable, to change length and width), and very high degrees of interleaving of memory. The NEC SX machines also have two additional features that are distinctive: The first is the use of branch prediction, which is rare for a vector-supercomputer. The second is the replication of all sets of functional units: if a functional unit of type X is replicated n times, then a single vector operation of that type is carried out by assinging vector element-pair number m to copy m **mod** n of the functional unit, thus effectively improving performance by a factor of n for that operation.

6.7 SUMMARY

This chapter has been an introduction to the design and programming of vector-pipe lined machines; the topics covered include fundamental concepts, placement and access of vectors in memory, vectorization, and performance issues. Of the two main types of vector architecture discussed, the register-to-register architecture has proved to be the better performer, overall, and is likely to remain the dominant design. Although the future of vector supercomputers continues to be the subject of debate, those computer manufacturers that have been leaders in this area — these are Cray, NEC, Hitachi, and Fujitsu — are still continuing with the development of new machines, either in single-processor systems or in multiple-processor systems. The only exception to this is CDC: both the parent company and the its spinoff company (ETA) are no longer in business.[4] There is also the possibility that the multimedia area will spawn new demand for (perhaps different types) of vector machines. Lastly, it is no longer the case that vector pipelines need be associated with just the large supercomputers: at least one manufacturer (Fujitsu) has introduced a single-chip vector-pipelined unit that is intended for use as a coprocessor with microprocessors [2]; according to the designers of this machine, for very large data sets, vector processing still gives better performance that the superscalar processing of conventional high-performance microprocessors.

---

[4]ETA went out of business in 1989, and with that CDC effectively got out of supercomputer development. In 1992 CDC itself was split into two information services companies.

7 INTERRUPTS AND BRANCH MISPREDICTIONS

This chapter deals with the handling of interrupts in pipelined machines and with recovery in the event of branch mispredictions. The first section is a discussion of basic implementation techniques, the second consists of a number of case studies, and the third is a summary.

In a non-pipelined machine, each instruction is completely processed before the next one is started, and when the need for an interrupt[1] occurs, both the faulting instruction and the machine-state just prior to the initiation of the faulting instruction — that is, the state that would exist when all instructions that logically precede the faulting one have had their intended effect on the state and all other instructions have had no effect — can be precisely identified; this is a *precise interrupt*, the obverse of which is an *imprecise interrupt*. In a pipelined machine, on the other hand, it is not always easy to know what instruction caused an interrupt condition or to have a well-defined machine-state (i.e. one that corresponds to non-pipelined sequential processing), unless special measures are taken: First, it may not be easy to associate an instruction with a fault, because two or more instructions in the pipeline simultaneously cause interrupts; for example, an instruction in the Decode stage might cause an illegal-opcode interrupt condition at the same time as another requires a page-fault interrupt in the Operand-Access stage and a third causes an error

---

[1]Sometimes a distinction is made between *interrupt* and *exception*; we shall shall not do this.

in the Execute stage. Some decision has to be made on a processing order in such a case. Second, it might be the case that by the time the need for an interrupt is recognised, the faulting instruction is long past the point at which it caused the interrupt condition; for example, in the CDC Cyber 205 there is one instance (the "data-flag-branch") where an interrupt requirement may go unrecognised until thirty-five instructions later. Third, machine-state can be changed by different instructions in progress at the same time, and it may be difficult to determine precisely the state as it would exist if each instructions were run to completion one at a time; moreover, this difficulty is aggravated if execution is out-of-order, as an instruction might cause an interrupt condition after an instruction that logically follows it has been executed. Consequently, interrupts have historically been problematic in pipelined machines.

Beyond the three main issues indicated above, there are also basic performance issues to be taken into account: Consider, for example, a simple linear pipeline that issues and executes instructions in strict program-order. In order to have a well-defined state, when an interrupt is recognized, all instructions for which the program counter has been altered must be allowed to complete. The delay incurred depends on the number of stages between the stage with the control adder and the end of the pipeline, and it can be high in some cases. Not surprisingly, therefore, the decision has been made in the design of some machines not to have precise interrupts if doing so would affect performance.

We have also seen in Chapter 4 that in a number of current high-performance machines the problem of ensuring a sustained flow of instructions in the face of branch instructions is (partially) solved by predicting the direction that a branch will take and then speculatively processing the instructions on the predicted path until the branch instruction is finally executed. If such prediction is incorrect, then the machine-state (which most likely will have been changed by the speculatively processed instructions) must be restored to what it was at the (logical-order) time that the processing of the branch instruction was initiated. The situation here is similar to part of the interrupt problem, except that the faulting instruction here can be more readily identified; accordingly, the same techniques can be, and have been, used to deal with both problems. We shall therefore also discuss branch misprediction in part of what follows. It should, however, be noted that some simplification is possible with respect to branches, as branch conditions must be resolved before the first completed instruction on the predicted path reaches the point where it would change architectural state.

Not all interrupts need to be processed with the same precision: There are those (e.g. page-faults) that need to be processed right away and that must be precise (or at least appear to be), since the faulting instruction cannot otherwise continue; but there are others (e.g. parity error) that do not have such requirements. We can also identify cases (e.g. arithmetic errors) where, in principle, a precise interrupt is required but in which imprecision can be tolerated. In the past some machine designs (e.g. the Cray supercomputers) have taken advantage of this and, giving performance the higher priority, settled for imprecise interrupts in the case of arithmetic and similar errors. Nevertheless, this is rapidly changing. There are at least two main reasons for the new state of affairs: the first is that since the techniques required for recovery with branch misprediction can also be used for interrupts, the cost of ensuring interrupt-precision latter can largely be amortized; the second is that the new IEEE standard on floating-point arithmetic has requirements that make imprecise interruption no longer acceptable for arithmetic operations. And, of course, users are simply not as tolerant as they used to be. As a result, almost all of the recent high-performance machines implement precise interrupts, even with out-of-order execution, and in this chapter, we shall look at the means of doing this. One notable exception to this new state of affairs is the DEC Alpha: the architecture provides no guarantees of precise interrupts in the case of arithmetic operations, on the grounds that doing so would compromise the performance of implementations; nevertheless, although the first implementations of this architecture did not provide for precise interrupts, the latest one (the Alpha 21264) does [7].

7.1   IMPLEMENTATION TECHNIQUES FOR PRECISE INTERRUPTS

Solutions to the problem of ensuring precise interrupts generally fall into two categories, according to how they maintain consistency with the sequential-execution model: in one category, instructions change architectural machine-state only in program order, thus ensuring that the state is always well defined; in the other, instructions change the state in the order in which they execute, but facilities exist to reconstruct a well-defined state in the event of an interrupt. There are five well-known implementation techniques, and a derived sixth, that have been suggested for the implementation of precise interrupts, and we shall describe them below as originally presented by their designers [5, 6, 15, 16, 18]. A number of case studies are then given to show the use of these techniques in practical machines.

The difficulty that arises because two or more instructions in the pipeline can cause interrupts at the same time can be resolved by giving priority to instructions that have made the most progress (i.e. those farthest along in the pipeline), the type of interrupt, and so forth; we shall not concern ourselves with this. We shall also initially assume that the architectural machine-state consists of just the architectural registers and that all instructions are scalar; after the primary discussions, we shall then consider exceptions to these assumptions. Following the terminology introduced in Chapter 1, we shall use *completion* to refer to when an instruction leaves an execution unit and *retire* to when an instruction has its intended effect on the machine-state. The goal, then, is to ensure in-order retirement, even if completion is out-of-order.

7.1.1   In-order completion

This is the simplest possible solution: an instruction is dispatched/issued, and allowed to modify machine-state, only if no preceding instruction has caused

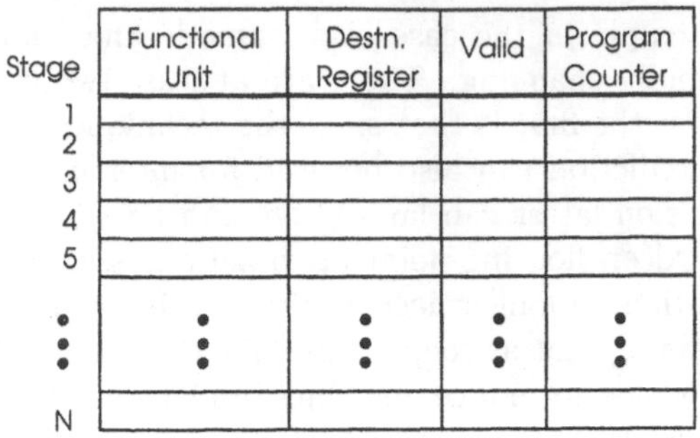

Stage	Functional Unit	Destn. Register	Valid	Program Counter
1				
2				
3				
4				
5				
⋮	⋮	⋮	⋮	⋮
N				

Figure 7.1.   Result shift register

an interrupt condition [15]. That is, by forcing completion to be in-order, it is ensured that retirement will be in-order.

The corresponding implementation is as follows. A *result shift register* is used to control the writing of results. The register has as many lines as there are stages (cycles) in the longest execution pipeline, and each line consists of four fields (Figure 7.1): a functional unit number, the address of a destination register, a program counter value, and a bit indicating whether or not the contents of the line are valid. When an instruction that requires n cycles for execution is issued, an attempt is made to make an entry in line n of the result shift register. If that line or any of lines $1 \ldots n - 1$ is use, i.e. has a set validity-bit, then the attempt is repeated on every cycle until it is successful. When an entry can be made, it consists of the identity of the functional unit that will execute the instruction, the destination-register address for the result, the instruction's program-counter value, and a set validity bit; at the same time all lines $1 \ldots n - 1$ that are not in use are filled with null information but made "valid". The latter prevents a following short-latency instruction from reserving any of the lines preceding n, and therefore ensures that no instruction is issued if it will finish execution before a logically preceding instruction.

The result shift register is then used to ensure precise interrupts as follows. Entries are processed in first-in/first-out order. On every cycle, the controls for the writing of the results are taken from the next line of the register if the corresponding instruction has not caused an interrupt condition; if an interrupt condition occurred, then the control logic for the result-writing cancels all pending writes into the register file. Since each entry in the result shift register is processed only after preceding entries, the register file is modified only as indicated by program-order and therefore has well-defined state when an interrupt is required.

The main drawback of this system is that it prevents a full use of multiple function units, as some units will sometimes have to be idle just to ensure in-order completion: blocked instructions can unnecessarily hold up following instructions that would otherwise be eligible for execution, and in particular, short-latency instructions are unnecessarily disadvantaged if they follow long-latency ones but there are no hazards to prevent them from otherwise being

executed. If it is possible to determine before an instruction has been executed
that it will not cause an interrupt or if the time when it can cause an interrupt
can be precisely determined, then the use of the result shift register may be
modified so that when an entry is made, only those entries that correspond
to finishing times before the time that an interrupt would be recognized are
"nullified"; however, this may not always be possible and is a partial solution
at best.

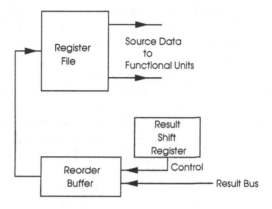

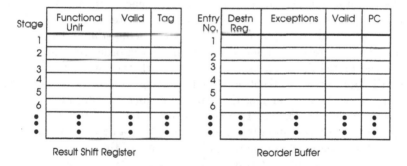

Figure 7.2.   Reorder buffer implementation

7.1.2   Reorder-buffer

A *reorder buffer* is one structure that allows instructions to complete out-of-
order but modify machine-state in program-order and is probably the most
commonly implemented of the systems described here [15]. The essential idea
here is to use some temporary storage for results computed out-of-order and
then commit these results to the register file in program-order. The buffer is
a circular first-in/first-out queue and is used in conjunction with a result shift
register; a complete arrangement is shown in Figure 7.2. The result shift register here no longer holds the result-destination address or the program-counter

value, as both now appear in the reorder buffer. The result shift register also has a new field — the Tag — which is the address of the reorder-buffer entry for the corresponding instruction. In the reorder buffer, one pointer, Tail, indicates the slot for the next instruction dispatched/issued, and another pointer, Head, indicates the next instruction to modify machine-state. In each line of the buffer, the Result (eventually) holds the result from the corresponding instruction, the Exceptions indicates what interrupt conditions occur during the execution of the instruction, and the other fields correspond to the result-destination address and the program counter in the original result shift register. The number of lines in the result shift register is still determined by the slowest operation, but that of the reorder buffer depends only on the desired performance (and cost): instruction dispatch/issue must stop if the buffer is full, so the buffer size determines the maximum number of instructions that can be active concurrently.

The reorder-buffer system is used as follows. When an instruction is dispatched/issued, it is allocated an entry in both the result shift register and the reorder buffer; and when it completes, the Tag from the result shift register is used to locate the reorder-buffer entry, and the results and any interrupt conditions are then written in. Each entry in the reorder buffer is processed when it reaches the head of the queue. At that time the Exceptions field is examined, and if there is no recorded need for an interrupt, then the data in the Result field is written into the register file. Thus if a particular instruction requires an interrupt, then by the time its entry reaches the head of the queue, all preceding instructions will have modified the register file, in the right order, leaving a state that corresponds to that of strict sequential execution.

The use of the reorder-buffer is an advance on in-order-completion, in that it has none of the disadvantages associated with the latter, but it may unnecessarily hold up an instruction that is waiting to issue if operands can be read only from the register file; this can be avoided by allowing results to go directly from the reorder buffer to the execution units, but at the cost of additional logic — comparators to detect the need for a bypass and multiplexors to realize the bypass paths (Figure 7.3). A system of this type is implemented in the AMD K5 (Section 7.2.3) as well as in a number of other machines. In some such implementations, the result shift register and reorder buffer are essentially combined into a single unit, by, for example, using associative search to locate entries. Note, however, that this implies that the number of lines is not necessarily optimal, but it results in some simplifications in the design and control.

7.1.3 History-buffer

The *history buffer* solves the problem (in the reorder buffer) of instructions being held up even though their operands have been computed, but this is done without without the need to implement all the bypass logic implied above [15]. The essential idea is to allow instructions to modify the machine-state as they finish but to retain enough state-history so that interrupts, when they occur,

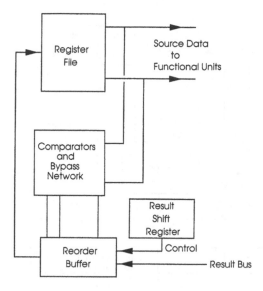

Figure 7.3.   Reorder buffer with bypass

can be precise. The organization for the use of the history buffer is shown in Figure 7.4. The result shift register is the same as in the reorder-buffer system, and the history buffer itself is similar to the reorder buffer, except that the Value field now stores the *old* value (i.e. the value to be overwritten with the instruction's result) of the named destination register. The history buffer may therefore be viewed as a reorder buffer that is used in a slightly different way from that above, and the size constraints are as for the latter. As with the reorder buffer, the two structures of Figure 7.4 may be combined into a single one; and in a simple pipeline the basic idea may be restricted to only a few registers of the machine-state, e.g. to the instruction-fetch and processor-status registers.

The history-buffer system is used as follows. When an instruction is dispatched/issued, it is allocated the next free entries in the result register file and the history buffer, the current value of the result-destination register is copied into the the Old-Value field of the history-buffer entry, and the other fields of the two structures are filled in as appropriate. And when the instruction completes, the result is written into the register file, and any need for an interrupt is recorded in the history buffer. An entry that reaches the head of the queue without requiring an interrupt is discarded, as the corresponding instruction must have finished successfully. If on the other hand, the Exceptions field for that instruction indicates that an interrupt is required, then the Value field of all valid buffer entries are written, from most recent to oldest, into the register file, thus reconstructing the state to what would it have been in program-order completion, and the program counter is reset from the entry of the history buffer.

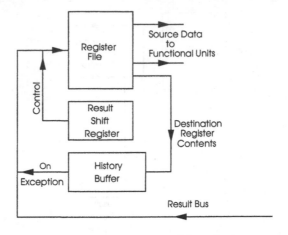

Figure 7.4.   History buffer implementation

Other than in the method of use, the main difference between the reorder buffer and the history buffer is in implications for the implementation of the register-file and on performance: First, the register file now requires an extra read port to get the old value at the same time as the operands are read out; the same ports may be used, but with an effect on performance. Second, the reconstruction of state, by writing a few register values at a time, can adversely affect performance.

The MIPS R10000 (Section 7.2.4) is an example of a recent machine that employs (what is essentially) a history buffer. In this implementation, the collection of physical registers that at any given moment are not mapped by the renaming (see Chapter 5) hold a partial state-history and so are, in essence, used as the data field of the history buffer: the mechanism in this case involves using the Active List to restore register old mappings.

7.1.4   Future-file

The *future-file* system is similar to the history-buffer system, in that it is based on saving a partial history of the machine-state, but it has none of the draw-backs of the latter [15]. The basic idea is maintain two register files such that one, the *architectural register file*[2], reflects the state in the sequential-execution model, and the other, the *future file*, serves as temporary storage that is modified by instructions as soon as they finish execution, i.e. in any order. (It is therefore the future file, rather than the architectural file, that is used as the working space for the functional units.) The two files are used in conjunction with a reorder buffer, as shown in Figure 7.5. Results are written into the reorder buffer at the same time as they are in the future file and are processed as they are with the reorder-buffer system, i.e. the results are written from the reorder buffer to the register file in program-order. Thus when an interrupt occurs, the architectural file is always well-defined and is used to restore the contents of the future file. The Toshiba TX3 is an example — the only one that we know of — of a machine that implements the future-file system [13].

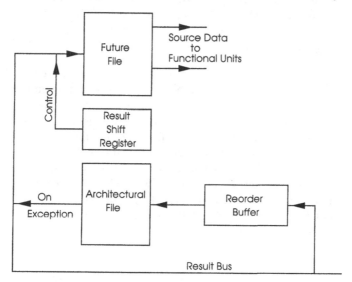

Figure 7.5.   Future file implementation

A major drawback of the future-file system is that it requires a duplication of the registers that hold the machine-state, and this can be quite large. There is, however, at least one case where complete duplication is not required: If renaming is implemented with a physical register file that is larger than the architectural register file, then at any given moment the unmapped physical registers will hold a partial recent history of the machine state and may be used as the architectural file; the situation is therefore similar to that mentioned above in the MIPS R10000. As with the history buffer, the basic idea may be implemented in a limited fashion, depending on the simplicity of the pipeline design.

---

[2]In this chapter we shall use *architectural registers* to refer to the physical registers that directly correspond to the registers specified in the architecture.

7.1.5 Checkpoint-repair

Checkpointing refers to the saving of machine-state at particular stages (*checkpoints*) of execution; and *repair* refers to the restoration of such state. A checkpoint can be made at a fine (instruction-by-instruction) level or at a coarse level (between several instructions); as the granularity increases, cost decreases (since less state must be saved), but so too does the precision of saved state [6]. Checkpoint-repair may therefore be viewed as a generalization of the above systems that record partial state history. For the implementation, there are basically two approaches: one is to copy at each checkpoint the contents of all the storage that implement the architectural name spaces; the other is to incrementally store at each checkpoint only the *differences* in state. The latter implementation is easily extendible to a modified history buffer. An example of an implementation of checkpoint-repair is described in Section 7.2.6.

7.1.6 Register update unit

The *register update unit* is a structure that has been suggested for the implementation of precise interrupts in combination with renaming and the detection and resolution of hazards; in essence, it combines a renaming table, reservation stations, and reorder buffer [16]. The unit consists of a first-in/first-out queue in which each entry (for a single instruction) has the format shown in Figure 7.6: the reservation-station part has a Data field for each operand and a Tag field and Ready bit for each source operand; the Issued bit indicates whether or not the instruction has been forwarded to an execution unit, and if it has, the Functional Unit field identifies which unit; the Executed bit indicates whether or not the instruction has left the execution unit; and the last field is the instruction's program-counter value.

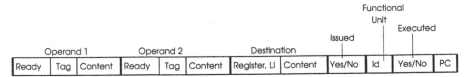

Figure 7.6.   Entry in register update unit

When an instruction is dispatched, it is assigned the next free slot in the register update unit; if necessary, instruction-dispatch is stalled until a free entry is available. This allocation, combined with an appropriate use of tags, accomplishes renaming. The mechanism used to coordinate data between producers and consumers is as follows. Each data register has two associated counters: the Number-of-Instances (NI) counter records the number of instances a register has been renamed (and, therefore, the number of times the register is named as a destination in the register update unit), and the Latest-Instance (LI) counter holds the number of the last such instance. Both counters are incremented when a renaming takes place (i.e. when an initial entry is made in the register update unit), and NI is decremented when a corresponding entry is deleted from the register update unit. (If NI is not zero, then there is an uncompleted instruction that is due to write into the register.) Whenever the register file is accessed for an operand, it returns either the contents (if valid) of the named register or a tag that consists of the register address and the LI value; the latter is used to ensure that following instructions get the latest contents of the register and that entries in the register update unit get the correct data.

Matching of data from execution units with instructions in the register update unit is done by a straightforward implementation of Tomasulo's algorithm (Chapter 5). Instructions with ready operands are issued from the register update unit in order of age, and as each instruction is issued, its Issued bit is set in order to avoid repeated issue. Retirement in an order that facilitates precise interrupts is exactly as in the reorder buffer: if the Executed bit is set for the entry at the head of the update queue — i.e. if the corresponding instruction has been executed without the need for an interrupt — then the data in the Destination field is written into the register file and the entry deleted. As with the reorder buffer system, the register update unit may be implemented with or without bypass paths.

7.1.7   Other aspects of machine-state

So far we have assumed that the state consists of just the architectural register file, including the program counter. But in most machines there will be other registers, such as the processor status word and condition-code registers, that are also architecturally visible parts of the state; and in some cases it may also be necessary to save part of the architecturally-invisible state. Caches, memory operations, virtual memory, and vector operations also need to be considered [15].

The same techniques used above for computational instructions can be used with most other instructions that change the machine-state. For example, for a memory STORE operation, assuming no cache, a dummy entry can be made in the structure used (result shift register, reorder buffer, or history buffer) and then used to control the memory access unit. When the dummy entry reaches the head of the queue it sends a signal to the memory access unit to perform the operation, which is stalled (in, say, a STORE or WRITE buffer) until then. This ensures that the memory state is always correctly defined when an interrupt occurs. Condition codes may be treated similarly by making suitable entries for instructions that set condition codes and treating the conditions as data; and the same applies to other processor status registers that are architecturally visible.

For caches, the system used depends on the write policy implemented. One straightforward solution is to treat the cache as a register file and make appropriate entries in the structures used for retirement-ordering, with cache-line address or data in place of register address or data. This solution can be used with both write-through and copy-back cache, although in the latter there is one case that requires special handling: with a reorder buffer, whenever a line is to be written back into memory, a check must be made to determine if the buffer has any data that is yet to be written into that line, and if there is, then the write-back must be stalled until the data has been written from the buffer into the cache; with a history buffer, either the cache line must be saved in the history buffer, or the write-back must be stalled until the corresponding entry reaches the end of the buffer queue. Another solution that can be used with a write-through cache is to immediately write into the cache but still process the actual memory operation in the manner indicated above. In this case the state of the cache may not always be consistent with that of memory, and the cache must be flushed when an interrupt occurs.

Virtual memory is one area where precise interrupts are crucial, since it is not possible for execution to continue until the interrupt has been serviced and the faulting instruction restarted. Assuming a Load-Store architecture, the structures above can be used to ensure that such interrupts are processed properly by arranging it so that for LOAD and STORE instructions, all virtual addresses are translated in program-order and that entries in whichever of the structures above is used to realize precise interrupts are read before the corresponding instructions have been checked for interrupt status. When a virtual-memory interrupt is necessary, the entry for the faulting instructions, and all following LOAD or STORE instructions are deleted from the addressing pipeline. Alternatively, the machine-state may be frozen while processing the interrupt, an example of which is the solution used for virtual memory interrupts in the MU5 computer (which is not of a Load-Store architecture): a special queue is used to hold all instructions for which there are pending memory accesses, and when a page fault occurs, the operating system ideally services by using instructions (to manipulate the translation registers) that do not involve the queue; should it become necessary to change process, the contents of the queue are dumped into memory and later restored [11]. A simple pipeline design can also lessen the difficulties involved: for example, in the first MIPS implementation, the pipeline is organized so that any page fault is recognised before the instruction causing it can issue a WRITE to memory; and when a page fault occurs, any WRITEs (up to and including the faulting instruction) that are in the pipeline are cancelled, thus permitting the easy draining and restarting of the pipeline.

Precise interruption is, as might be expected, more problematic with a vector pipeline than with a scalar pipeline: in general, a single vector operation produces a result of many elements, each of which changes machine-state, and a vector operation can be interrupted midway in execution. There is therefore an implicit requirement for the saving and restoration of partially completed vectors, as well as a large amount of associated control information (such as field lengths), and a straightforward extension of the methods discussed above may require large replication of registers and other buffers. The potential complexity can be appreciated by considering that in the CDC Star-100, which was originally intended to be an all-hardwired machine, the operations involved in the startup and rundown of vector operations during interrupts were so complex that it became necessary to microprogram them.[3] The system that has been in used in many vector-pipelined machines, such as the CDC Cybers and the Crays, is derived from the *exchange-package* method of the CDC 6600 [2]. The essential idea is that when an interrupt occurs, the contents of all buffers, control registers, partial results, etc. that are required to restart the faulting instruction from the point of interruption are saved in an *exchange package* (or *invisible package*) and restored from there when the interrupt has been serviced. (Similar implementations will be found in some simple scalar pipelines that save the small amount of required restart-information in *shadow registers* that track the normal registers.) The CDC Cybers are further designed to lessen the difficulties involved by always maintaining a one-page lookahead for result vectors, to ensure that results already in the pipeline can be safely returned to memory, and disabling virtual addressing (i.e. taking all addresses to be real) during the processing of interrupts.

7.1.8 Summary

We have discussed the problem of precise interrupts in pipelined machines and given a number of solutions that have been adopted. In what follows we shall look at some case studies that exemplify the use in practical machines of the various techniques described above. Branch mispredictions can also be dealt with using these techniques if such mispredictions are viewed as just another type of "interrupt": all that is necessary is to ensure that the misprediction is recognised by the time the entry for the first speculatively issued instruction reaches the head of the retirement-ordering structure used and that when a misprediction is detected, all following entries, and including, that for the first mispredicted instruction (as well as those in memory queues) are cleared; in this case the first speculatively dispatched/issued instruction is marked as requiring an interrupt. Alternatively, an arrangement may be made that includes making an entry for the branch instruction and retiring that entry only when the direction of the branch has been resolved. Then, with a system such as the reorder buffer or the history buffer, it is easy to ensure that speculatively processed instructions do not change machine-state and interrupts indicated by such instructions are not processed until the resolution of the branch on which they depend confirms that the instructions were on the correct path. A slightly different approach is used in the MIPS R10000 (Section 7.2.5), in which the most of the relevant machine state is saved separately for each branch instruction.

At least one study has been carried out to evaluate, in terms of cost and performance of a VLSI implementation, most of the systems discussed above [20]. The results of this are summarized in Table 7.1 for the model machine of Figure 7.7. The conclusions are approximately as we should expect: the

---

[3]The microprogramming proved to be so beneficial that in later CDC machines, such as the Cyber 205, microprogramming is used to control all aspects of vector operations.

in-order-completion system costs the least, but has the worst performance, and the history file and future file have the best performance, with the latter as the more costly; in terms of cost-to-performance ratio, the history-file system is the best. In general, however, the best choice depends on the particular pipeline implementation — on such factors as the pipeline length, size of the relevant architectural state, the method use to implement renaming, and so forth: for example, the history buffer is likely to be slower than the reorder buffer if there are few register-file ports and there can be many completed but unretired instructions. In practice the reorder buffer is the most often implemented method because it is easy to also use teh same mechanism for hazard-detection and renaming. Other studies of cost and performance issues have shown that certain simplifications that reduce costs and increase performance are possible if some imprecision is selectively allowed [12].

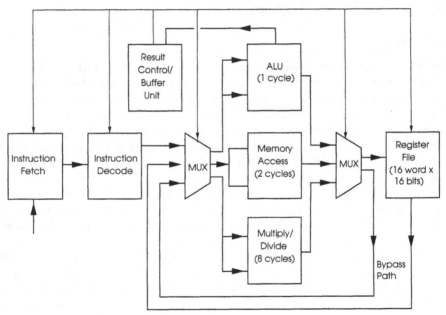

Figure 7.7.   Model for evaluation of precise-interrupt systems

The systems above may be implemented fully as described, or the essential idea may only be partial implemented. An example of a partial implementation is in the MU5 computer. The machine uses branch prediction and has its control point five stages into the pipeline, but one of the main addressing registers (the stack pointer) can be changed at the first stage. If a prediction is wrong, then all of the instructions behind the control point must be discarded and, unless special action is taken, the value in that register will not be correct.[4]   The

---

[4]A similar problem with stack addressing also occurs in the National Semiconductor NS32532/32580; see Section 7.2.7.

special action taken consists of having each instruction carry along with it the old value of the register, and on a misprediction the value associated with the instruction at the control point is used to restore the contents of the register (see Section 4.4.6.3). Evidently, this is no more than a simple approach to saving history, and in a linear pipeline it can be extended to other registers as well.

Table 7.1.   Evaluation of precise-interrupt systems

Method	Structure [bits/entry]	Cost*	Perf**	Cost:
Imprecise	Result shift register [4]	1.000	1.00	1.00
In-order	Result shift register [20]	1.042	0.81	1.29
Reorder buffer	Result shift register [6]	1.148	0.83	1.38
	Reorder buffer [38]			
History buffer	Result shift register [11]	1.148	0.85	1.35
	History buffer [37]			
Future file	Result shift register [6]	1.457	0.85	1.71

*normalized size, **normalized

7.2   CASE STUDIES

The following case studies show practical applications of some of the techniques discussed above: the PowerPC 604 and the AMD K5 use variations of the reorder buffer; the Metaflow uses a variant of the register update unit; the Motorola 88110, the National Semiconductor NS32532/32580, and the MIPS R10000 use history buffers; and the SPARC64 uses checkpoint repair. With one exception, all of these machines implement renaming, and the precise-interrupt technique used in each such case is closely connected to the renaming system used; the reader may therefore find it useful to first review the relevant sections of Chapter 5 before proceeding. The case studies also cover branch misprediction.

7.2.1 Motorola 88110

The Motorola 88110 processor uses a single mechanism to handle mispredicted branches and interrupts: a history buffer is used to record all of the architectural machine-state as each instruction is issued [4, 19]. Up to two instructions can be issued in each cycle, and reconstruction of state is also carried out at a similar rate; issue is in-order, bt completion can be out-of-order. The algorithm to retire entries from the history buffer is run at the end of each cycle, and in each run as many entries as can be retired are; so the status of the buffer can change from "full" to "empty" in a single cycle.

Whenever an instruction causes a synchronous (internal) interrupt, e.g. a page fault or an arithmetic error, all of the instructions that were issued before it are allowed to run to completion, and any LOAD instructions that logically follow the faulting instruction and have been granted access to the cache or the bus are allowed to complete but are prohibited from writing their results. Once the logically preceding instructions have been completed, any instructions (except those accessing the cache or the bus) that were issued after the faulting one are flushed from the pipeline, and the history buffer is used to rollback the machine state to what it was just before the faulting instruction was issued. For STORE instructions, each instruction has an entry made in the history buffer at the time the instruction is issued to the memory Load/Store unit, but no update of memory takes place until the instruction is retired from the buffer. A timing diagram for the sequence of operations in an internal interrupt is given in Figure 7.8: the interrupt condition occurs in the interval A–B; outstanding memory operations complete in the interval B–C, whose length depends on the number of such instructions; the machine-state is restored in the interval C–D, whose length (up to six cycles) depends on the number (up to eleven) of instructions issued after the faulting one; and the transfer of control to the interrupt handler takes place in the three-cycle interval D–E.

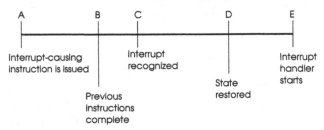

Figure 7.8.   Timing of internal interrupt for MC88110

Asynchronous (external) interrupts are are of two modes — *normal* and *windowed* — and either can be maskable or non-maskable. The handling of such interrupts is largely similar to that for internal interrupts, except for the differences arising from the fact that an external interrupt is not necessarily associated with an instruction in the history buffer. A non-maskable windowed interrupt is recognized and processed after stopping instruction-issue and allowing for the completion of all instructions issued before the assertion of the interrupt; for a maskable interrupt, on the other hand, instruction-issue continues up to the next branch instruction, and then all issued instructions are allowed to complete before the interrupt condition is recognised. In normal-interrupt mode, the main goal is to minimize response time without affecting performance. The sequence of operations involved is as follows (Figure 7.9). The interrupt is received in the processor two cycles (for a maskable interrupt) or three cycles (for a non-maskable interrupt) after it is asserted and detected after all memory operations involving the cache or the bus are complete. Once the interrupt has been detected, all remaining memory operations are tagged as requiring an interrupt, and instructions leaving the execution units are similarly tagged in the cycle after which they write their results; the tagging takes place in the interval from when the interrupt is detected to when the first tagged instruction reaches the top of the buffer and the interrupt is recognised. If the history buffer is empty, then the first instruction that would been issued after the detection of the interrupt enters the history buffer and is tagged as requiring an unimplemented-opcode interrupt. Once the interrupt is recognized, the sequence of operations is as for an internal interrupt: the state is restored, the pipeline is flushed, and the interrupt handler is run.

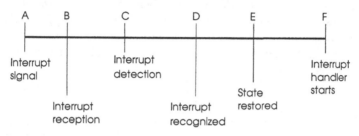

Figure 7.9.   Timing of external interrupt for MC88110

For branches, when a misprediction is detected, the first instruction in the predicted path is tagged as having an unimplemented opcode, and recovery is made in the same way as for an internal interrupt. To ensure a precise memory state, whenever there is a pending predicted branch, the memory does not service STOREs or cache misses on LOADs until the direction of the branch they are associated with has been resolved.

7.2.2 PowerPC 604

The mechanism used to ensure interrupt precision in the PowerPC 604 is essentially a reorder buffer that does not store any data; the data is held in a separate structure [17]. A 16-line buffer is used to keep track of all instructions in progress as well as their status. As each instruction is dispatched, it is allocated an entry in the buffer, which can process up to four instructions in each cycle. An instruction completes only when it is at the head of the queue and it requires no interrupt. The reorder buffer is used in conjunction with renaming (see Section 5.4.6), and a separate buffer (the Rename Buffer) holds an instruction's result prior to its being committed into the register file; results may be written into the Rename Buffer out of program-order, but the reorder buffer ensures that they are committed in the right order.

The description in Section 7.1 of the use of reorder and history buffers assumes that when an instruction is at a point where it has been executed without interruption and all logically preceding instructions have completed, a write-back of its results immediately takes place. In a superscalar machine this requires at least as many write-ports on the register file as the maximum number of instructions that can complete in a single cycle — and perhaps even more. In the PowerPC 604, this difficulty is dealt with by decoupling the the completion stage from the write-back stage: actual writing is done at the Write-back stage or at the Completion stage, depending on the availability of write ports. This reduces the number of write-ports required from the eight that it would otherwise be to two.

Branch mispredictions are handled as follows. When one occurs, all instructions on the mispredicted path are nullified wherever they are — in the reservation stations, execution units, and memory queues — and corresponding temporary results in the Rename Buffer are discarded. Some instructions, such as those that operate on Condition-code registers are, however, deemed to be too expensive to execute out-of-order and are *serialized*. This serialization process involves issuing such an instruction to the execution unit but with a tag indicating that it should not be executed until all prior instructions have completed successfully.

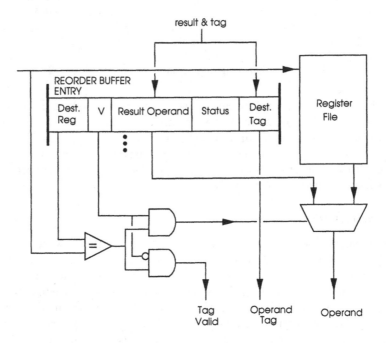

Figure 7.10.   AMD K5 reorder buffer entry

7.2.3 AMD K5

The AMD K5 uses a reorder buffer for both interrupts and branch misprediction [3]. Each decoded instruction is assigned a buffer entry that has the format shown in Figure 7.10; the various fields are the address of the destination register for the instruction, a slot for the result, some status bits that record interrupt and other conditions, a unique tag for the result, and a validity bit. The reorder buffer is also used for renaming, so different instructions with the same architectural destination register will have buffer entries that have the same Destination Register entry but different Destination Tags; this aspect of the buffer's use is discussed in detail in Section 5.4.7.

An instruction receives an operand from the reorder buffer (if it is there) or from the register file: an operand is in the reorder buffer if there is a line that has a set validity bit and whose Destination Register matches the instruction's source-operand-register address. There is therefore no need to wait for the result to be committed. Two types of stalling may occur in the use of the buffer: if it is full, then decoding is temporarily suspended; and if an entry is at the head of the queue but its result has not yet been returned, then it is held there until the data is available, but decoding and the allocation of new buffer entries continue as long as there is space.

The need for an interrupt is recognised when an entry reaches the head of the queue and the Exceptions fields so indicates; branch mispredictions, on the other hand, are recognised by the buffer as soon the corresponding branch instruction is executed. An interrupt requirement is acted upon immediately or at later time, depending on its nature. Such action consists of emptying the reorder buffer, aborting all uncompleted instructions in the pipeline, and running a microcoded interrupt-handling routine. For a branch misprediction, the action taken consists of setting a "cancel" bit in the Status field of every buffer entry that follows that for the branch instruction. The functional units continue to execute instructions on the mispredicted path, and to write results into the buffer, but all "cancelled" entries are simply discarded as they reach the head of the reorder queue. This cancellation method results in a more efficient use of logic than in a system (such as that of the PowerPC 604 or MIPS R10000) in which instructions to be aborted must first be tracked down throughout the machine.

7.2.4 MIPS R10000

The system used for precise interrupts in the MIPS R10000 is essentially a history buffer in which each entry has been modified to store the address of a temporary register that holds a result instead of storing the result itself [22]. The main structure used is a 32-line circular queue, called the *active list*, that holds one entry for each instruction in progress, with allocations made in program-order. Because the processor can issue up to four instructions in each cycle, to ensure that Active-List entries can be made at a matching rate, the list is implemented as four 8-entry queues. The Active List is also used in conjunction with renaming (Section 5.4.5). Each entry of the list has the format shown in Figure 7.11: the Logical Destination is the address of the architectural destination-register for the instruction, the Old Destination is the address of the last physical register to which that architectural register was mapped, the Done bit indicates whether or not the corresponding instruction has completed execution, and the Exceptions and Condition-Codes fields are self-explanatory.

Each dispatched instruction carries an identifying 5-bit tag that is the address of its entry in the Active List; after the instruction has been completed, the tag is used to locate the Active-List entry, the various fields of which are then filled in as appropriate. When an entry reaches the head of the list without the corresponding instruction requiring an interrupt, the entry is discarded, the old physical register is returned to the list of free registers, and the architectural register thereafter considered mapped to the last physical register associated with it. If, on the other hand, an interrupt-condition occurred, then the entry is simply deleted and the Old-Destination values in the Active List used to restore the mapping table used for renaming to its correct state; this leaves the architectural state as it was prior to the dispatching of the faulting instruction. (The "unmapping" is done in reverse order — from the most recent entry to the oldest — as a single architectural register can be renamed more than once and can therefore have more than one entry in the Active List.)

Old Destination	Logical Destination	Done	Exceptions	Condition Codes

Figure 7.11.   MIPS R10000 active-list entry

The system used for recovery in the event of a branch misprediction is based on the assumption that such mispredictions occur more frequently than interrupts and, therefore, recovery must be much faster. Accordingly, to handle branch misprediction, the machine uses a four-entry *branch stack* that is used to save or restore state in a single step. When a branch instruction is decoded (and its direction predicted), it is assigned an entry in the stack, unless the stack is full, in which case decoding is stalled until a free entry becomes available. A Branch-Stack entry consists of the alternate target-address, copies of the renaming mapping tables, and various control and status bits. Subsequent instructions on the predicted path go through the pipeline accompanied by a 4-bit *branch mask* that indicates what entry of the stack they are associated with. When an outstanding branch-condition is finally resolved, if the prediction turns out to have been incorrect, then the corresponding Branch-Stack entry is used to restore state, and all active instructions with a matching Branch Mask are aborted. The mask bits on active instructions are reset as soon as the corresponding branch condition has been determined.

For memory operations, a 16-entry first-in/first-out queue (associated with the data cache) is used to hold the addresses of all pending operations: a LOAD or STORE instruction is allocated the next free entry in the queue when it is decoded, which entry is deleted when the instruction completes successfully. STORE operations are coordinated with the Active List: the processing of such an instruction (i.e. the writing of data into the cache) is carried out when the instruction's Active-List entry is at the head of that queue. In the event of a branch misprediction, all entries in the address queue that were made after the dispatching of the branch instruction are deleted.

7.2.5 Metaflow Lightning

The Metaflow Lightning uses a single structure, the Deferred-scheduling Register-renaming Instruction Shelf (DRIS) as a renaming table, reservation stations, and reorder buffer — in effect, an implementation of the register update unit [14]. (The first two aspects of the DRIS are discussed in detail in Section 5.4.8.) The format of each DRIS entry is shown in Figure 7.12. The use of the DRIS as a reorder buffer is straightforward: instructions are entered in program order and retired if they complete without interruption. We shall therefore discuss only the branch misprediction aspects.

To handle branches, condition codes are treated as implicitly addressed destination registers, and branch-instruction entries in the DRIS are therefore similar to those for other instructions; this ensures that the condition-code register, which is part of the machine-state, is always well-defined. If at issue-time the condition on which a branch depends is not known, then the instruction's DRIS entry is "locked" until the the result of the condition-evaluation is available; processing continues on a predicted path, and the DRIS entry records the direction of the prediction. Whenever an instruction that sets a condition code completes, its UID is compared with the Source UIDs of all locked (branch) instructions, and those entries that yield a match are unlocked. The oldest branch instruction is then issued for execution, and on completion its taken direction is compared with the predicted direction recorded in the DRIS. If the two match, then no action is required; otherwise all speculatively processed instructions are aborted — by deleting all subsequent entries in the DRIS.

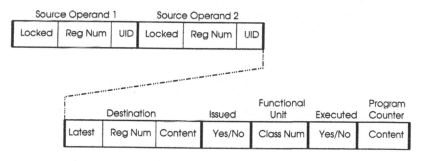

Figure 7.12.   Entry in Metaflow DRIS

7.2.6 SPARC64

The SPARC64 implements checkpoint-repair and uses this for both interrupts and branch misprediction [21]. The completion of instructions is carried out in three pipeline stages, after execution: the Deactivate stage determines whether or not an instruction has completed without requiring an interruption or if a rollback is required to restore state, the Commit stage writes temporary results into the register file, and the Retire stage frees resources associated with instructions whose results have been committed.

The processor maintains a circular first-in/first-out queue, the Active Ring, that records the status (*dispatched, waiting, executing, completed,* or *committed*) of all instructions in progress; up to sixty-four instructions can be active at any given time, and each is identified by a unique 6-bit number. Three pointers are associated with the Active Ring: one that points to the entry of the last instruction dispatched, one that points to the last instruction whose results were committed, and one that points to the last instruction retired. The results of an instruction are committed only if it and all preceding instructions (i.e. those with preceding entries in the Active Ring) have completed successfully.

In order to realize precise recovery in the event of an interrupt or branch misprediction, the machine has a two-level recovery process: *Backups* restore the machine-state to the last checkpointed state: each instruction, as it proceeds, is accompanied by a pointer to such a state. And, since the checkpoints are made between several instructions, and these restorations therefore only approximate the exact state required, *backsteps* allow the restoration process to be completed on an instruction-by-instruction basis, with corresponding changes made to the program counter. The checkpoints are made for branches and other instructions that can modify non-renamed processor control state. Each checkpoint consists of the renaming map tables and the contents of those registers that are not renamed at the time of checkpointing. Up to 16 checkpoints can be maintained (within register files), which means there can be up to 16 unresolved branches in progress at a given moment, and a single cycle suffices to restore state to a checkpoint. Backsteps, on the other hand, are carried out incrementally, at a rate of up to 4 instructions per cycle.

Branch instructions are processed by two dedicated units: the Branch Unit (which also predicts the directions of branches) and the Branch Misprediction Unit. To handle branch misprediction, a specialized store is also provided to hold the contents of the branch prediction table when a prediction is made; if the prediction is incorrect, a restoration is made from this store. For instructions that are already waiting to execute, each reservation station also includes the last checkpoint number at the time the instruction was dispatched; in the event of a misprediction, the checkpoint number is sent to the reservation stations, where instructions with a matching number are aborted. To further facilitate fast recovery in the event of a misprediction, the alternate address of a branch instruction is held in a special register.

7.2.7 National Semiconductor NS32532/32580

We stated in the introduction that a major motivation for the implementation of precise-interrupt mechanisms has been the requirement in the IEEE standard for floating-point arithmetic that exceptions be precise. We now look at a case study exemplifying this: the National Semiconductor NS32532/32580 [9]. This processing system consists of two units: a 32-bit general-purpose microprocessor (the NS32532) and a floating-point arithmetic unit composed of a controller (the NS32580) and computational subunit (a Weitek WTL3164). The precise-interrupt mechanism used is essentially a history buffer.

The organization of the machine, with respect to the recovery mechanisms, is shown in Figure 7.13. The basic protocol for interaction between the two units is that the CPU when it decodes a floating-point instruction issues it to the floating-point controller (FPC) and continues processing other instructions; the FPC eventually responds with an indication of whether or not the instruction was completed successfully or whether an interrupt is required. The history buffer is realized as a a set of *shadow registers* that hold the old values associated with destination registers of instructions in progress and a Floating-Point Instruction FIFO Buffer (FIF) that holds, in proper order, the program-counter (PC) and program-status-register (PSR) values associated with these instructions. When an instruction is issued to the the FPC, the PC-PSR values at the time are entered into a new line of the FIF and the value in any named destination register is written into a shadow register. Responses from the FPC are in an order that matches the FIF entries, and when the CPU receives a response, either an entry is deleted from the FIF (successful completion of an instruction) or (if an interrupt is required) the FIF and shadow registers are used to restore the PC, PSR, and architectural registers to their states at the time that the instruction with the leading FIF entry was issued; in the latter case, the FIF is also cleared of other entries.

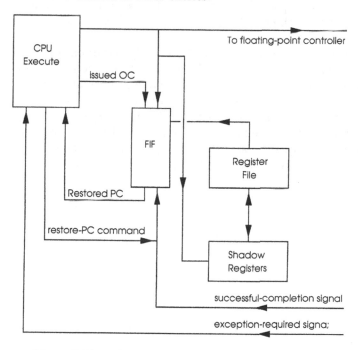

Figure 7.13.   Interrupt mechanism in NS32532/32580

There are three exceptions to the above processing sequence; these are those cases in which a floating-point instruction has a memory location for its destination, modifies the the PSR, or uses a top-of-stack addressing mode. In the first case, the CPU issues the instruction to the FPC and then stops until the FIF is empty (i.e. until all previously issued instructions have been completed without interrupts); this ensures that updates to memory never occur out of order. In the second case, the CPU also stops issuing further instructions to the FPC until the instruction that modifies the PSR completes successfully and the FPC requests CPU intervention to update the PSR. In the third case, the difficulty is that instructions that use stack addressing can implicitly change the value in the stack-pointer register. The basic problem could have been solved by extending the FIF to also hold stack-pointer values, but the designers of the machine decided that the extra complications could not be justified by the small performance gains to be got from pipelining the processing of the few floating-point instructions that use top-of-stack operands. Therefore, the implemented solution is simply to not issue such an instructions until the FIF becomes empty.

7.3 SUMMARY

We have discussed the problem in a pipelined machine of realizing precise interrupts and precise recovery in branch misprediction, and we have also described a number of techniques that can be used as solutions — even with out-of-order-execution machines. A number of practical case studies have also been given.

In the past the implementation of precise interruption was considered not worthwhile, with respect to cost and performance, except in very few cases (e.g. page faults). This has now changed as more machines implement superscalar processing, aggressive branch prediction techniques that require recovery similar to an interrupt, and implement the IEEE standard on arithmetic (with its requirement for precise interrupts to handle arithmetic errors). We should therefore except that the use of the techniques discussed here, as well as newer ones, will be widespread in future machines.

BIBLIOGRAPHY

Chapter 1

1. AMD. 1997. AMD-K6 MMX Processor. Advanced Micro Devices, Sunnyvale, California.

2. Buckle, J. 1978. *The ICL 2900 Series*. McMillan, London, United Kingdom.

3. Chen, T.C. 1970. Overlap and pipeline processing. In: H.S. Stone, Ed., *Introduction to Computer Architecture* (SRA, Chicago), Chapter 9.

4. Chen, T.C. 1971. Parallelism, pipelining, and computer efficiency. *IEEE Computer* (January):69–74.

5. Cragon, H.C. 1996. *Memory Systems and Pipelined Processors*. Jones and Bartlett Publishers, Boston, MA.

6. Cragon, H.C. and W.J. Watson. 1989. The TI Advanced Scientific Computer. *IEEE Computer*, 22(1):55–65.

7. Deverell, J. 1975. Pipeline iterative arithmetic arrays. *IEEE Transactions on Computers*, C-24: 317–322.

8. Diep, A.T., C. Nelson, and J.P. Shen. 1995. Performance evaluation of the PowerPC 620 microarchitecture. In: *Proceedings, 23rd International Symposium on Computer Architecture*, pp 163–174.

9. Durdan, W.H., W.J. Bowhill, J.F. Brown, W.V. Herrick, R.M. Marcello, S. Samudrala, G.M. Uhler, and N. Wade. 1990. An overview of the VAX 6000 Model 400 chip set. *Digital Technical Journal*, 2(2):36–51.

10. Edmondson, J.H., P. Rubinfield, R. Preston, and V. Rajagopalan. 1995. Superscalar instruction execution in the Alpha 21164 microprocessor. *IEEE Micro*, 15(2):33–43.

11. Flynn, M.J. 1995. *Computer Architecture: Pipelined and Parallel Processor Design*. Jones and Bartlett Publishers, Boston

12. Gwenap, L. 1996. Digital 21264 sets new standard. *Microprocessor Report*, 10(14).

13. Hennessy, J.L. and D.A. Patterson. 1990. *Computer Architecture: A Quantitative Approach*. Morgan Kaufmann Publishers, San Mateo, California.

14. Hennessy, J., N. Jouppi, and J. Gill. 1981. MIPS: A VLSI processor. In: H.T. Kung, B. Sproull, and G. Steele (Eds.), *VLSI Systems and Computations* (Computer Science Press, Rockville, Maryland), pp 347–346.

15. Hockney, R.W. and C.R. Jesshope. 1988. *Parallel Computers* (Adam Hilger, Bristol, United Kingdom), Chapter 2.

16. Hwang, K. and F. Briggs. 1984. *Computer Architecture and Parallel Processing* (McGraw-Hill, New York), Chapter 3.

17. Ibbett, R.N. and N.P. Topham. 1989. *The Architecture of High Performance Computers* (Springer-Verlag, New York), volume 1, Chapter 4.

18. Intel. 1997. Intel Architecture Software Developer's Manual – Volume 1: Basic Architecture. Intel Corporation, Mt. Prospect, Illinois.

19. Jouppi, N.P. 1989. The non-uniform distribution of instruction-level and machine parallelism and its effect on performance. IEEE *Computer*, 38(12):1645–1658.

20. Jouppi, N.P. and D.W. Wall. 1989. Available instruction-level parallelism for superscalar and superpipelined machines. In: *Proceedings, 3rd International Conference on Architectural Support for Programming Languages and Operating Systems*, pp 272–282.

21. Keller, R.M. 1975. Look-ahead processors. *Computing Surveys*, 7(4):177–195.

22. Kogge, P.M. 1981. *The Architecture of Pipelined Computers*. McGraw-Hill, New York.

23. McMahon, S.C., M. Bluhm, and R.A. Garibay. 1995. 6x86: the Cyrix solution to executing x86 binaries on a high-performance microprocessor. *Proceedings of the IEEE*, 83(12):1664–1672.

24. MIPS. 1995. MIPS R10000 Microprocessor User's Manual. MIPS Technologies, Mountain View, California.

25. Mirapuri, S., M. Woodacre, and N. Vasseghi. 1992. The MIPS R4000 processor. *IEEE Micro*, 12(4):10-22.

26. Morris, D. and R.N. Ibbett. 1979. *The MU5 Computer System*. Springer-Verlag, New York.

27. Omondi, A. R. 1994. *Computer Arithmetic Systems: Algorithms, Architectures, and Implementations*. Prentice-Hall International, London, United Kingdom.

28. Ramamoorthy, C.V. and H.F. Li. 1977. Pipeline architecture. *Computing Surveys* 9(1):61–102.

29. Sites, R. L. 1992. Alpha AXP architecture. *Digital Technical Journal*, 4(4):19–35.

30. Smith, J.E., and G.S. Sohi. 1995. The microarchitecture of superscalar processors. *Proceedings of the IEEE*, 83(12):1609–1621.

31. Song, S.P., M. Denman, and J. Chang. 1994. The PowerPC 604 microprocessor. *IEEE Micro*, 14(10):8–17.

32. Song, P. 1997. IBM's Power3 to replace P2SC. *Microprocessor Report*, 11(5).

33. Sun Microsystems. 1995. The UltraSPARC Processor:Technology White Paper. Mountain View, California.

34. Topham, N.P., A. Omondi, and R.N. Ibbett. 1988. On the design and performance of conventional pipelined architectures. *Journal of Supercomputing*, 1(4):353–393.

35. Watson, W.J. 1972. The TI-ASC — a highly modular and flexible super computer architecture. In: *Proceedings, AFIPPS Fall Joint Computer Conference*, pp 221–228.

36. Williams, T., N. Paktar, and G. Shen. 1995. SPARC64: A 64-b 64-active-instruction out-of-order-execution MCM processor. *IEEE Journal of Solid-State Circuits*, 30(11):1215–1226.

37. Yeager, K.C. 1996. The MIPS R10000 superscalar microprocessor. *IEEE Micro*, 16(2):28–40.

Chapter 2

1. Bowhill, W.J. et al. 1995. Circuit implementation of a 300MHz 64-bit second-generation CMOS Alpha CPU. *Digital Technical Jurnal*, 7(1):100–118.

2. Cotten, L.W. 1965. Circuit implementation of high-speed pipeline systems. In: *Proceedings, AFIPS Fall Joint Computer Conference*, pp 489–504.

3. Cotten, L.W. 1969. Maximum-rate pipeline systems. In: *Proceedings, AFIPS Spring Joint Computer Conference*, pp 581–586.

4. Dobberpuhl, D.W. et al. 1992. A 200MHz 64-bit dual-issue CMOS microprocessor. *Digital Technical Journal*, 4(4):35–50.

5. Davidson, E.S. et al. 1975. Effective control for pipelined computers. In: *Proceedings, IEEE COMPCON*, pp 181–184.

6. Donchin, D.R. et al. 1992. The NVAX CPU chip: design challenges, methods, and CAD tools. *Digital Technical Journal*, 4(3):24–37.

7. Dubey, P.K. and M.J. Flynn. 1990. Optimal pipelining. *Journal of Parallel and Distributed Computing*, 8(1):pp 10–19.

8. Earle, J.G. 1965. Latched carry-save adder. *IBM Technical Disclosure Bulletin*, 7:pp 909–910.

9. Fawcett, B.K. 1975. Maximum clocking rates for pipelined digital systems. M.S. thesis, Department of Electrical Engineering, University of Illinois, Urbana-Champaign, Illinois, USA.

10. Ganguley. S., D. Lehther, and S. Pullela. 1997. Clock distribution methodology for PowerPC microprocessors. *Journal of VLSI Signal Processing*, 16(2-3):181–189.

11. Klas, F. and M.J. Flynn. 1993. Comparative studies of pipelined circuits. Technical report No. CSL-TR-93-579, Computer Systems Laboratory, Stanford University, California.

12. Kunkel, S.R. and J.E. Smith. 1986. Optimal pipelining in supercomputers. In: *Proceedings, 13th International Symposium on Computer Architecture*, pp 404–411.

13. Hallin, T.G. and M.J. Flynn. 1972. Pipelining of arithmetic functions. *IEEE Transactions on Computers*, C-21:880-886.

14. Mead, C. and L. Conway. 1980. *Introduction to VLSI Systems*. Addison-Wesley, Reading, Massachusetts.

15. Morris, D. and R.N. Ibbett. 1979. *The MU5 Computer System*. Springer-Verlag, New York.

16. Samaras, W.A. 1987. The CPU clock system in the VAX 8800 family. *Digital Technical Journal*, 4:34–40.

17. Unger, S.H. and C.-J. Tan. 1986. Clocking schemes for high-speed digital systems. *IEEE Transactions on Computers*, C-35(10):880–895.

18. Weste, N.H.E. and K. Eshragian. 1993. *Principles of CMOS VLSI Design*. Addison-Wesley, Reading, Massachusetts.

19. Yuen, J. and C. Svensson. 1989. High-speed CMOS circuit techniques. *Journal of Solid-State Circuits*, 24(1).

Chapter 3

1. AMD. 1997. AMD-K6 MMX Processor. Advanced Micro Devices, Sunnyvale, California.

2. Alpert, D.B. and M.J. Flynn. 1988. Performance trade-offs for microprocessor cache memories. *IEEE Micro*, 8(4):44-54.

3. Boland, L.J., G.D. Granito, A.U. Marcotte, B.U. Messina, and J.W. Smith. 1967. The IBM System/360 model 91: storage system. *IBM Journal of Research and Development*, 11(1):54-68.

4. Budnik, P.P. and D.J. Kuck. 1971. The organization and use of parallel memories. *IEEE Transactions on Computers*, C-20(12):1566-1569.

5. Burnett, G.J. and E.G. Coffman. 1975. Analysis of interleaved memory systems using blockage buffers. *Communications of the ACM*, 18(2):91-95.

6. CDC 1987. Cyber 200 Model 205 Computer System. Hardware Reference Manual. Control Data Corporation, Minneapolis, Minnesota, USA.

7. Diefendorff, K. and M. Allen, 1992. Organization of the Motorola 88110 super-scalar RISC microprocessor. *IEEE Micro*, 12(4):40-63.

8. Edmondson, J.H., P. Rubinfield, R. Preston, and V. Rajagopalan. 1995. Superscalar instruction execution in the Alpha 21164 microprocessor. *IEEE Micro*, 15(2):33-43.

9. Fite, D.B., T. Fossum, and D. Manley. 1990. Design strategies for the VAX 9000 system. *Digital Technical Journal*, 2(4):13-24.

10. Hill, M.D. 1987. Aspects of Cache Memory and Instruction Buffer Performance. Technical Report No. UCB/CSD 87/382, Computer Science Division, University of California, Berkeley.

11. Hill, M.D. 1988. A case for direct-mapped caches. *IEEE Computer*, December.

12. Hsu, W.-C. and J.E. Smith. 1998. A performance study of cache prefetching methods. *IEEE Transactions on Computers*, 47(5):497-508.

13. Inayoshi et al. 1988. Realization of the Gmicro/200. *IEEE Micro*, 8(2):12-21.

14. Jouppi, N. 1990. Improving direct-mapped cache performance by addition of a small fully-associative cache and prefetch buffers. In: *17th International Symposium on Computer Architecture*, pp 364-373.

15. Jouppi, N.P. 1993. Cache write policies and performance. In: *Proceedings, 20th International Symposium on Computer Architecture*, pp 191-201.

16. Jouppi, N.P. and S.J.E. Wilton. 1994. Trade-offs in two-level on-chip caching. In: *Proceedings, 21st International Symposium on Computer Architecture*, pp 34-45.

17. Kroft, D. 1981. Lockup-free instruction fetch/prefetch cache organization. In: *Proceedings, 8th International Symposium on Computer Architecture*, pp 81-85.

18. Lawrie, D., and C.Vora. 1982. The prime memory system for array access. *IEEE Transactions on Computers*, 31(5):435-432.

19. Matick, R, R. Mao, and S. Ray. 1989. Architecture, design, and operating characteristics of a 12ns CMOS functional chip cache. *IBM Journal of Research and Development*, 33(5):524-539.

20. Matick, R.E. 1977. *Computer Storage Systems and Technology*. Wiley and Sons, New York.

21. Meade, R.M. 1971. Design approaches for cache memory control. *Computer Design*, January.

22. Morris, D. and R.N. Ibbett. 1979. *The MU5 Computer System*. Springer-Verlag, New York.

23. Pohm, A.V. and O.P. Agrawal. 1983. *High-Speed memory Systems*. Reston Publishers, Reston, Virginia, USA.

24. Przybylski, S.A. 1990. *Cache Design: A Performance-Directed Approach*. Mogran Kaufmann Publishers, San Mateo, California.

25. Rau, R., D.W.L. Yau, W. Yen, and R.A. Towle. 1989. The Cydra 5 departmental supercomputer. *IEEE Computer*, 22(1):12–35.

26. Rau, B.R. 1991. Pseudo-randomly interleaved memory. In: *Proceedings, 18th International Symposium on Computer Architecture*, pp 74–83.

27. Smith, A.J. 1978. Sequential-program prefetching in memory hierarchies. *IEEE Computer*, December.

28. Smith, A.J. 1987. Line (block) size for CPU cache hierarchies. *IEEE Transactions on Computers*, C-36(9):1063–1075.

29. Smith, A.J. 1982. Cache memories. *ACM Computing Surveys*, 14(3):473-530.

30. Song, P. 1997. IBM's Power3 to replace P2SC. *Microprocessor Report*, 11(5).

31. Tse, J. and A.J. Smith. 1998. CPU prefetching: timing evaluation of hardware implementations *IEEE Transactions on Computers*, 47(5):509–526.

32. Thornton, J.E. 1970. *Design of a Computer: The Control Data 6600*. Scott, Foresman, and Co., Illinois, USA.

33. Tremblay, M. , D. Greenley, and K. Normoyle. 1995. The design of the microarchitecture of the UltraSPARC-1. *Proceedings of the IEEE*, 83(12).

Chapter 4

1. Alpert, D. and D. Avnon. 1993. Architecture of the Pentium microprocessor. *IEEE Micro*, June: 11–21.

2. Alsup, M. 1990. Motorola's 88000 family architecture. *IEEE Micro*, 10(3):48–66.

3. AMD 1987. Am29000 Streamlined Instruction Processor: User's Manual. Advanced Micro Devices, Sunnyvale, California, USA.

4. AMD 1997. AMD-K6 MMX Processor. Advanced Micro Devices, Sunnyvale, California.

5. Asprey, T. et al. 1993. Performance features of the PA7100 microprocessor. *IEEE Micro*, 13(3):22–35.

6. Ball, T. and J.R. Larus. 1993. Branch prediction for free. In: *Proceedings, ACM SIGPLAN Conference on Programming Language Design and Implementation*.

7. Becker, M.C. et al. 1993. The PowerPC 601 microprocessor. *IEEE Micro*, 13(5):54–68.

8. Blickenstein, D.S. et al. 1992. The GEM optimizing compiler system. *Digital Technical Journal*, 4(4):121–136.

9. Bray, B. and M.J. Flynn. 1991. Strategies for branch target buffers. In: *Proceedings, 24th Workshop on Microprogramming and Microarchitecture*, pp 42–49.

10. Calder B. and D. Grunwald. 1994a. Reducing branch costs via branch alignment. In: *Proceedings, 6th International Symposium on Architectural Support for Programming Languages and Operating Systems*, pp 242–251.

11. Calder B. and D. Grunwald. 1994b. Fast and accurate branch prediction. In: *Proceedings, 21st Annual International Symposium on Computer Architecture*, pp 2–11.

12. Calder B., D. Grunwald, and J. Elmer. 1995. A system level perspective on branch architecture performance. In: *Proceedings, 28th International Symposium on Microarchitecture*, pp 199–206.

13. Calder, B. and D. Grunwald. 1995. Next cache line and set prediction. In: *Proceedings, 22nd International Symposium Computer Architecture*, pp 287–296.

14. CDC 1975. Control Data 7600 Series and Cyber 70/Model 76 Computer Systems: Hardware Reference Manual. Control Data Corporation, Minneapolis, Minnesota, USA.

15. Chang, P.-Y., E. Hao, T. Yeh, and Y.N. Patt. 1994. Branch classification: a new mechanism for improving branch prediction performance. In: *Proceedings, 27th International Symposium on Microarchitecture*.

16. Chang, P.-Y., E. Hao, and Y.N. Patt. 1995. Alternative implementations of hybrid branch predictors. In: *Proceedings, 28th International Symposium on Microarchitecture*, pp 252–257.

17. Chang, P.-Y., E. Hao, and Y.N. Patt. 1995. Target prediction for indirect jumps. In: *Proceedings, 22nd International Symposium Computer Architecture*, pp 274–283.

18. Chang, P.-Y., M. Evers, and Y. Patt. 1996. Improving branch prediction accuracy by reducing pattern history table interference. In: *Proceedings, International Conference on Parallel Architectures and Compilation Techniques*.

19. Chow, P. and M. Horowitz. 1987. Architectural tradeoffs in the design of the MIPS-X. In: *Proceedings, 14th International Symposium on Computer Architecture*, pp 300–308.

20. Christie, D. 1996. Developing the AMD K5 architecture. *IEEE Micro*, 16(2):16–26.

21. Circello, J. et al. 1995. The superscalar architecture of the MC68060. *IEEE Micro*, 15(2):10–21.

22. Cortadella, J. and T. Jove. 1988. Designing a branch target buffer for executing branches with zero times cost in a RISC processor. *Microprocessing and Microprogramming*, 24:573–580.

23. Cragon, H.G. 1992. *Branch Strategy Taxonomy and Performance Models*. IEEE Computer Society Press, Los Alamitos, California.

24. CRAY. 1989. Cray-1 Computer Systems: Mainframe Reference Manual. Cray Research, Inc., Mendota Heights, Minnesota.

25. Davidson, J.W. and D.B. Whalley. 1990. Reducing the cost of branches by using registers. In: *Proceedings, 17th Annual International Symposium on Computer Architecture*, pp 182–191.

26. DeRosa, J. and H. Levy. 1987. An evaluation of branch architectures. In: *Proceedings, 14th Annual International Symposium on Computer Architecture*, pp 10–16.

27. Diefendorff, K. and M. Allen, 1992. Organization of the Motorola 88110 superscalar RISC microprocessor. *IEEE Micro*, 12(4):40–63.

28. Diep, T.A., C. Nelson, and J.P. Shen. 1995. Performance evaluation of the PowerPC 620 microarchitecture. In: *Proceedings, 22nd Annual International Symposium on Computer Architecture*, pp 163–174.

29. Ditzel, D.R. and H.R. McLellan. 1987. Branch folding in the CRISP microprocessor: reducing branch delay to zero. In: *Proceedings, 14th Annual International Symposium on Computer Architecture*, pp 2–9.

30. Driesen, K. and U. Holzle. 1998. Accurate indirect branch prediction. *Proceedings, 25th Annual International Symposium on Computer Architecture*, pp 167–178.

31. Dubey, P.K. and M.J. Flynn. 1991. Branch strategies: modeling and optimization. *IEEE Transactions on Computers*, 40(10):1159–1167.

32. Dutta, S. and M. Franklin. 1995. Control flow prediction with tree-like subgraph for superscalar processors. In: *Proceedings, 28th International Symposium on Microarchitecture*, pp 258–263.

33. Eden, A.N. and T. Mudge. 1998. The YAGS branch prediction scheme. In: *Proceedings, 31st International Symposium on Microarchitecture*.

34. Edmondson, J.H. et al. 1995. Internal organization of the Alpha 21164, a 300-MHz 64-bit quad-issue CMOS RISC microprocessor. *Digital Technical Journal*, 7(1):119–135.

35. Edmondson, J.H., P. Rubinfield, R. Preston, and V. Rajagopalan. 1995. Superscalar instruction execution in the Alpha 21164 microprocessor. *IEEE Micro*, 15(2):33–43.

36. Emma, P.G. and E.S. Davidson. 1987. Characterization of branch and data dependencies for evaluating pipeline performance. *IEEE Transactions on Computers*, 36(7):859–876.

37. Emer, J.S. and D.W. Clark. 1984. A characterization of processor performance in the VAX-11/780. In: *Proceedings, 11th Annual International Symposium on Computer Architecture*, pp 301–309.

38. Evers, M., S. Patel, R. Cappel, and Y. Patt. 1998. What makes two-level branch prediction work. *Proceedings, 25th Annual International Symposium on Computer Architecture*, pp 52–61.

39. Fagin, B. and A. Mital. 1995. The performance of counter- and correlation-based schemes for branch target buffers. *IEEE Transactions on Computers*, 42(12):1383–1393.

40. Fagin, B. and R. Russell. 1995. Partial resolution in branch target buffers. In: *Proceedings, 28th International Symposium on Microarchitecture*, pp 193–198.

41. Farrens, M.K. and A.R. Pleszkun. 1994. Implementation of the PIPE processor. *IEEE Computer*, 24(1):65–71.

42. Fisher, J.A. and S.M. Feudenberger. 1992. Predicting conditional branch directions from previous runs of a program. In: *Proceedings, 5th International Conference on Architectural Support for Programming Languages and Operating Systems*, pp 85–95.

43. Gerosa, G. et al. 1997. A 250MHz 5W PowerPC microprocessor with on-chip L2 cache controller. *IEEE Journal of Solid-State Circuits*, 32(11):1635–1649.

44. Gloy, N., M.D. Smith, and C. Young. 1995. Performance issues in correlated branch prediction schemes. In: *Proceedings, 28th International Symposium on Microarchitecture*, pp 3–14.

45. Gonzalez, A.M. and J.M. Llaberia. 1993. Reducing branch delay to zero in pipelined processors. *IEEE Transactions on Computers*, 42(3):363–371.

46. Gonzalez, A.M., J.M. Llaberia, and J. Cortadella. 1988. A mechanism for reducing the cost of branches in RISC architectures. *Microprocessing and Microprogramming*, 24:565–572.

47. Grohoski, G.F. 1990. Machine organization of the IBM RISC System/6000 processor. *IBM Journal of Research and Development*, 43(1):37–58.

48. Gross, T.-L. and J.L. Hennessy. 1982. Optimizing delayed branches. In: *Proceedings: 15th Annual Workshop on Microprogramming*, pp 114–120.

49. Gwenap, L. 1997. Centaur improves C6 with no extra cost. *Microprocessor Report*, 11(15).

50. Gwenap, L. 1996. Digital 21264 sets new standard. *Microprocessor Report*, 10(14).

51. Halfill, T.R. Beyond Pentium II. *Byte*, Dec 1997.

52. Halfhill, T.R. 1996. AMD K6 takes on Intel P6. *Byte*, January:67–72.

53. R.H. Halstead, G.R. Gao, R.A. Iannucci, and B. Smith, Editors. 1994. *Multithreaded Computer Architecture: A Summary of the State of the Art*. Kluwer Academic Publishers, Boston, Massachusetts.

54. Hitachi 1998. Series Overview: SuperH RISC Engine Embedded Processors. Hitachi, Japan.

55. Hsu, P. Y.-T. 1994. Designing the TFP microprocessor *IEEE Micro*, 14(2):23–33.

56. Holgate and Ibbett. 1980. An analysis of instruction-fetching strategies in pipelined computers. *IEEE Transactions on Computers*, C-29(4):325–329.

57. Hwu, W.M., T.M. Conte, and P.P. Chang. 1989. Comparing software and hardware schemes for reducing the cost of branches. In: *Proceedings, 16th Annual International Symposium on Computer Architecture*, pp 224–233.

58. Iacobovici, S. 1988. A pipelined interface for high floating-point performance with precise exceptions. *IEEE Micro*, 8(3):77–87.

59. Ibbett, R.N. and N.P. Topham. 1989. *The Architecture of High Performance Computers* (Springer-Verlag, New York), volume 1, Chapter 4.

60. Inayoshi et al. 1988. Realization of the Gmicro/200. *IEEE Micro*, 8(2):12–21.

61. Jouppi, N.P. and D.W. Wall. 1989. Available instruction-level parallelism for superscalar and superpipelined machines. In: *Proceedings, 3rd International Conference on Architectural Support for Programming Languages and Operating Systems*, pp 272–282.

62. Juan, T., S. Sanjeevan, and J. Navarro. 1998. Dynamic history-length fitting: a third level of adaptivity for branch prediction. *Proceedings, 25th Annual International Symposium on Computer Architecture*, pp 155–166.

63. Kaeli, R. and P.G. Emma. 1991. Branch history table predictions of moving branch targets due to subroutine returns. In: *Proceedings, 18th Annual International Symposium on Computer Architecture*, pp 34–41.

64. Kanenko, H. et al. 1990. Realizing the V80 and its system support functions. *IEEE Micro*, April:56–59.

65. Katvenis, M. 1985. *Reduced Instruction Set Architecture for VLSI*. MIT Press, Boston, Massachusetts.

66. Katvenis, M. and N. Tzartzanis. 1991. Reducing the branch penalty by rearranging instructions in a double-width memory. In: *Proceedings, 4th International Conference on Architectural Support for Programming Languages and Operating Systems*, pp 15–27.

67. Kuiran, L. et al. 1991. Classification and performance on instruction buffering techniques. In: *Proceedings, 18th Annual International Symposium on Computer Architecture*, pp 150-159.

68. Lee, C.-C., I.-C. K. Chen, and T.N. Mudge. 1997. The bi-mode branch predictor. In: *Proceedings, 30th International Symposium on Microarchitecture*.

69. Lee, J.F.K. and A.J. Smith. 1984. Branch prediction strategies and branch target buffer design. IEEE *Computer*, 17(1):6–22.

70. Lee, R. 1989. Precision architecture. IEEE *Computer*, 22(1):78–91.

71. Lewis, D. K., J.P. Costello, and D.M. O'Connor. 1988. Design tradeoffs for a 40 MIPS (peak) CMOS 32-bit microprocessor. In: *Proceedings, International Conference on Computer Design*, pp 110–113.

72. Lilja, D.J. 1988. Reducing branch penalty in pipelined processors. *IEEE Computer*, 21(7):47–55.

73. Mahlke, S.A. et al. 1995. A comparison of full and partial predicated execution support for ILP processors. In: *Proceedings, 22nd Annual International Symposium on Computer Architecture*, pp 138–150.

74. Mahlke, S.A. et al. 1994. Charaterizing the impact of predicated execution on branch prediction. In: *Proceedings, 27th Annual International Symposium on Microarchitecture*, pp 217–227.

75. McFarling, S. 1993. Combining branch predictors. WRL Technical Note TN-36, Western Research Laboratory, Digital Equipment Corporation, Palo Alto California.

76. McFarling, S. and J. Hennessy. 1986. Reducing the cost of branches. In: *Proceedings, 13th Annual International Symposium on Computer Architecture*, pp 396–403.

77. McMahan, S.C., M. Bluhm, amd R.A. Garibay. 1995. 6x86: the Cyrix solution to executing x86 binaries on a high performance microprocessor. *Proceedings of the IEEE*, 83(12):1664–1672.

78. McLellan, E. 1993. The Alpha AXP architecture and 21064 microprocessor. *IEEE Micro*, 13(3):36–47.

79. Melear, C. 1989. The design of the 88000 RISC family. *IEEE Micro*, 9(2):26–38.

80. MIPS. 1995. MIPS R10000 Microprocessor User's Manual. MIPS Technologies, Mt. View, California. Michaud, P., A. Seznec, and R. Uhlig. 1995. Trading

conflict and capacity aliasing in conditional branch predictors. In: *Proceedings, 22nd Annual International Symposium on Computer Architecture*, pp 292–303.

81. Mirapuri, S., M. Woodacre, and N. Vasseghi. 1992. The MIPS R4000 processor. *IEEE Micro*, April: 10–22.

82. Miyata, M. et al. 1988. The TX1 32-bit microprocessor: performance analysis and debugging support. *IEEE Micro*, 8(2):37–46.

83. Morris, D. and R.N. Ibbett. 1979. *The MU5 Computer System*. Springer-Verlag, New York.

84. Murray, J.E., R.C. Hetherington, and R.M. Salett. 1990. VAX instructions that illustrate the architectural features of the VAX 9000 CPU. *Digital Technical Journal*, vol. 2, no. 4, pp 13–24.

85. Nair, R. 1995. Optimal 2-bit branch predictors. *IEEE Transactions on Computers*, 44(5):698–702.

86. Nair, R. 1995. Dynamic path-based branch correlation. In: *Proceedings, 28th International Symposium on Microarchitecture*, pp 15–23.

87. Oehler, R.R. and R.D. Groves. 1990. IBM RISC System/6000. *IBM Journal of Research and Development*, 43(1):23–36.

88. Okamoto et al 1988. Design considerations for 32-bit microprocessor TX3. In: *Digest of Papers, COMPCON*, pp 25–29.

89. Pan, S.T., K. So, and J.T. Rahmeh. 1996. Improving the accuracy of dynamic branch prediction using branch correlation. In: *Proceedings, 5th International Conference on Architectural Support for Programming Languages and Operating Systems*, pp 76–84.

90. Perlebeg, C.H. and A.J. Smith. 1993. Branch target buffer design and optimization. *IEEE Transactions on Computers*, 42(4):396–412.

91. Pnevmatikatas, D.N. and G.S. Sohi. 1994. Guarded execution and branch prediction in dynamic ILP processors. In: *Proceedings, 21st International Symposium on Computer Architecture*, pp 120–129.

92. Potter, M., M. Vaden, J. Young, and N. Ullah. 1994. Resolution of control and data dependencies in the PowerPC 601. *IEEE Micro*, 14(5):18–29.

93. Radin, G. 1983. The IBM 801 minicomputer. *IBM Journal of Research and Development*, 27(3):237–246.

94. Ramamoorthy, C.V. and H.F. Li. 1977. Pipeline architecture. *Computing Surveys* 9(1):61–102.

95. Rau, B.R. and G.E. Rossman. 1977. The effect of instruction fetch strategies upon the performance of pipelined instruction units. In: *Proceedings, 4th Annual International Symposium on Computer Architecture*, pp 80–89.

96. Rau, R., D.W.L. Yau, W. Yen, and R.A. Towle. 1989. The Cydra 5 departmental supercomputer. *IEEE Computer*, 22(1):12–35.

97. Russell, R.D. 1978. The PDP-11: A case study of how not to design condition codes. *Proceedings, 5th Annual International Symposium on Computer Architecture*, pp 190–194.

98. Sechrest, S., C.C. Lee, and T. Mudge. 1995. The role of adaptivity in two-level branch prediction. In: *Proceedings, 28th International Symposium on Microarchitecture*, pp 264–270.

99. Sechrest, S., C.C. Lee, and T. Mudge. 1996. Correlation and aliasing in dynamic branch predictors. In: *Proceedings, 23rd International Symposium on Computer Architecture*, pp 22–31.

100. Sequin, C.H. and D.A. Patterson. 1983. Design and implementation of RISC I. In: B. Randell and P.C. Treleaven, Eds., *VSLI Architecture* (Prentice-Hall International, U.K.), pp 276–298.

101. Sites, R.L. 1993. Alpha AXP Architecture. *Communications of the ACM*, 36(2):33–44.

102. Smith, J.E. 1981. A study of branch prediction strategies. In: *Proceedings, 8th Annual International Symposium on Computer Architecture*, pp 135–148.

103. Sprangle, E. et al. 1995. The Agree predictor: a mechanism for reducing negative branch history interference. In: *Proceedings, 22nd Annual International Symposium on Computer Architecture*, pp 284–291.

104. Sohie, G.R.L. and K.L. Kloker. 1988. A digital signal processor with IEEE floating-point arithmetic. *IEEE Micro*, 8(6):49–57.

105. Song, S.P., M. Denman, and J. Chang. 1994. The PowerPC 604 microprocessor. *IEEE Micro*, 14(5):8–17.

106. Srivastava, A. and A.M. Despain. 1993. Prophetic branches: a branch architecture for code compaction and efficient execution. In: *Proceedings, 26th International Symposium on Microarchitecture*.

107. Stark, J., M. Evers, and Y.N. Patt. 1998. Variable length path prediction. In: *Proceedings, 8th International Conference on Parallel Architectures and Compilation Techniques*

108. Su, C.-L. and A.M. Despain. 1994. Branch with masked squashing in super-pipelined processors. In: *Proceedings, 21st Annual International Symposium on Computer Architecture*, pp 130–140.

109. Talcott, A.r., M. Nemirovsky, and R.C. Wood. 1995. The influence of branch prediction table interference on branch prediction performance. In: *Proceedings, 3rd International Conference on Parallel Architectures and Compilation Techniques*.

110. Talcott, A.R. et al. 1994. The impact of unresolved branches on branch prediction performance. In: *Proceedings, 22nd Annual International Symposium on Computer Architecture*, pp 12–21.

111. Thornton, J.E. 1970. *Design of a Computer: the Control Data 6600*. Scott, Foresman, and Co.; Glenview, Illinois.

112. Topham, N.P., A. Omondi, and R.N. Ibbett. 1988. On the design and performance of conventional pipelined architectures. *Journal of Supercomputing*, 1(4):353–393.

113. Tremblay, M. , D. Greenley, and K. Normoyle. 1995. The design of the microarchitecure of the UltraSPARC-1. *Proceedings of the IEEE*, 83(12): 1653–1663.

114. Tyson, G. S. 1994. The effects of predicated execution on branch prediction. In: *Proceedings, 27th Annual International Symposium on Microarchitecture*, pp 196–206.

115. Uchiyama, K. et al. 1993. The Gmicro/500 superscalar microprocessor with branch buffers. *IEEE Micro*, 13(5)12–22.

116. Uht, A.K., V. Sindagi, and S. Somanathan. 1997. Branch effect reduction techniques. *IEEE Computer*, May: 71–80.

117. Vogel, J.P. and B.K. Holmer. 1994. Analysis of skip instructions in the HP Precision Architecture. In: *Proceedings, 27th Annual International Symposium on Microarchitecture*, pp 207–216.

118. Wilken, K.D. 1992. Toward zero-cost branches using instruction registers. In: *Proceedings, 25th International Symposium on Microarchitecture*, pp pp 214–217.

119. Williams, T., N. Patkar, and G. Shen. 1995. SPARC64: A 64-b 64-active-instruction out-of-order-execution MCM processor. *IEEE Journal of Solid-State Circuits*, 30(11):1215–1226.

120. Wu, Y. and J.R. Larus. 1994. Static branch frequency and program profile analysis. In: *Proceedings, 27th International Symposium on Microarchitecture*, pp 1–11.

121. Yeager, K.C. 1996. The MIPS R10000 superscalar microprocessor. *IEEE Micro*, 16(2):28–40.

122. Yeh, T.-Y. and Y.N. Patt. 1991. Two-level adaptive branch prediction. In: *Proceedings, 24th International Symposium on and Workshop on Microarchitecture*, pp 51–61.

123. Yeh, T.-Y. and Y.N. Patt. 1992. A comprehensive instruction fetch mechanism for a processor supporting speculative execution. In: *Proceedings, 25th International Symposium on and Workshop on Microarchitecture*, pp pp 129–139.

124. Yeh, T.-Y. and Y.N. Patt. 1992. Alternative implementations of two-level adaptive branch prediction. In: *Proceedings, 19th Annual International Symposium on Computer Architecture*, pp 124–134.

125. Yeh, T.-Y. and Y.N. Patt. 1993. A comparison of dynamic branch predictors that use two levels of branch history. In: *Proceedings, 20th Annual International Symposium on Computer Architecture*, pp 257–266.

126. Yeh, T.-Y. and Y.N. Patt. 1993. Branch history table indexing to prevent pipeline bubbles in wide-issue superscalar processors. In: *Proceedings, 26th Annual International Symposium on Computer Architecture*, pp 164–175.

127. Yeh, T.-Y., D.T. Marr, and Y.N. Patt. 1993. Increasing instruction fetch rates via multiple branch predictions and a branch address cache. In: *Proceedings, 7th Annual ACM International Conference on Supercomputing*, pp 67–76.

128. Yoshida, T. et al. 1992. The GMicro/100 32-bit microprocessor. *IEEE Micro*, August:62–72.

129. Young, H.C. and J.R. Goodman. 1984. A simulation study of architectural data queues and prepare-to-branch instruction. In, *Proceedings, International Conference on Computer Design*, pp 544–549.

130. Young, C. and M. Smith. 1994. Improving the accuracy of branch prediction using branch correlation. In: *Proceedings, 6th International Conference on Architectural Support for Programming Languages and Operating Systems*, pp 232–241.

131. Young, C., N.Gloy, and M.D. Smith. 1995. A comparative analysis of schemes for correlated branch prediction. In: *Proceedings, 22nd Annual International Symposium on Computer Architecture*, pp 276–286.

Chapter 5

1. Acosta, R.D., J. Kjelstrup, and H.C. Torng. 1986. An instruction issuing approach to enhancing performance in multiple functional unit processors. *IEEE Transactions on Computers*, C-35(9):815–828.

2. Alpert, D. and D. Avnon. 1993. Architecture of the Pentium microprocessor. *IEEE Micro*, 13(3):11–21.

3. AMD 1987. Am29000 Streamlined Instruction Processor: User's Manual. Advanced Micro Devices, Sunnyvale, California, USA.

4. AMD. 1997. AMD-K6 MMX Processor. Advanced Micro Devices, Sunnyvale, California.

5. Alsup, M. 1990. Motorola's 88000 family architecture. *IEEE Micro*, 10(3):48–66.

6. Asprey, T. et al. 1993. Performance features of the PA7100 microprocessor. *IEEE Micro*, 13(3):22–35.

7. Becker, M.C. et al. 1993. The PowerPC 601 microprocessor. *IEEE Micro*, 13(5):54–68.

8. Chen, T.C. 1964. The overlap design of the IBM System/360 Model 92 central processing unit. In: *Proceedings, AFIPS Fall Joint Computer Conference*, vol. 26, prt II, pp 73–80.

9. Christie, D. 1996. Developing the AMD-K5 architecture. *IEEE Micro*, 16(2):16–26.

10. Circello, J. et al. 1995. The superscalar architecture of the MC68060. *IEEE Micro*, 15(2):10–21.

11. Diefendorff, K. and M. Allen, 1992. Organization of the Motorola 88110 superscalar RISC microprocessor. *IEEE Micro*, 12(4):40–63.

12. Diep, T.A., C. Nelson, and J.P. Shen. 1995. Performance evaluation of the PowerPC 620 microarchitecture. In: *Proceedings, 22nd Annual International Symposium on Computer Architecture*, pp 163–174.

13. Edmondson, J.H. et al. 1995. Internal organization of the Alpha 21164, a 300-MHz 64-bit quad-issue CMOS RISC microprocessor. *Digital Technical Journal*, 7(1):119–135.

14. Edmondson, J.H., P. Rubinfield, R. Preston, and V. Rajagopalan. 1995. Superscalar instruction execution in the Alpha 21164 microprocessor. *IEEE Micro*, 15(2):33–43.

15. Grohoski, G.F. 1990. Machine organization of the IBM RISC System/6000 processor. *IBM Journal of Research and Development*, 36(1):37–58.

16. Keller, R.M. 1975. Look-ahead processor. *ACM Computing Surveys*, 7(4):66–72.

17. McMahon, S.C., M. Bluhm, and R.A. Garbay. 1995. 6x86: the Cyrix solution to executing x86 binaries on a high performance machine. *Proceedings of the IEEE*, 83(12): 1664–1672.

18. Melear, C. 1989. The design of the 8800 RISC family. *IEEE Micro*, 9(2):26–38.

19. MIPS. 1996. MIPS R10000 Microprocessor User's Manual. MIPS Technologies, Mt. View, California.

20. Morris, D. and R.N. Ibbett. 1979. *The MU5 Computer System*. Springer-Verlag, New York.

21. Omondi, A. R. 1994. Ideas for the design of multithreaded pipelines. In: R.H. Halstead, G.R. Gao, R.A. Iannucci, and B. Smith, Editors, *Multithreaded Computer Architecture: A Summary of the State of the Art*. Kluwer Academic Publishers, Boston.

22. Popescu, V. et al. 1991. The Metaflow architecture. *IEEE Micro*, June: 10–13, 63–73.

23. Potter, M., M. Vaden, J. Young, and N. Ullah. 1994. Resolution of control and data dependencies in the PowerPC 601. *IEEE Micro*, 14(5):18–29.

24. Sites, R.L. 1993. Alpha AXP Architecture. *Communications of the ACM*, 36(2):33–44.

25. Smith, J.E. 1989. Dynamic instruction scheduling and the Astronautics ZS-1. IEEE *Computer*, July:21–35

26. Smith, J.E., and G.S. Sohi. 1995. The microarchitecture of superscalar processors. *Proceedings of the IEEE*, 83(12): 1609–1624.

27. Song, S.P., M. Denman, and J. Chang. 1994. The PowerPC 604 microprocessor. *IEEE Micro*, 14(5):8–17.

28. Thornton, J.E. 1970. *Design of a Computer: The Control Data 6600*. Scott, Foresman, and Co., Illinois, USA.

29. Tomasulo, R.M. 1967. An efficient algorithm for exploiting multiple arithmetic units. *IBM Journal of Research and Development*, 11(1):25–33.

30. Tremblay, M. , D. Greenley, and K. Normoyle. 1995. The design of the microarchitecture of the UltraSPARC-1. *Proceedings of the IEEE*, 83(12): 1653–1663.

31. Trioani, M. et al. 1985. The VAX 8600 I Box: A pipelined implementation of the VAX architecture. *Digital Technical Journal*, 1:24–42.

32. Weiss, S. and J.E. Smith. 1984. Instruction issue logic in pipelined supercomputers. IEEE *Transactions on Computers*, vol. C-33, no. 9 (Sep. 1984), pp

33. Williams, T., N. Patkar, and G. Shen. 1995. SPARC64: A 64-b 64-active-instruction out-of-order-execution MCM processor. *IEEE Journal of Solid-State Circuits*, 30(11):1215–1226.

34. Yeager, K.C. 1996. The MIPS R10000 superscalar microprocessor. *IEEE Micro*, 16(2):28–40.

Chapter 6

1. Allen, R. and K. Kennedy. 1992. Vector register allocation. *IEEE Transactions on Computers*, 42(10):1290–1317.

2. Awaga, M. and H. Takahashi. 1993. The μVP 64-bit vector coprocessor. *IEEE Micro*, 13(5):24–36.

3. Baskett, F. and T.W. Keller. 1977. An evaluation of the Cray-1 computer. In: D.J. Kuck, D.H. Lawrie, and A.H. Sameh, Editors, *High-Speed Computer and Algorithm Organization* (Academic Press, New York), pp 71–84.

4. Bucher, I.Y. 1983. The computational speed of supercomputers. In: *Proceedings, ACM Sigmetrics Conference on Measurement and Modeling of Computer Systems*, pp 151–165.

5. Buchholz, W. 1986. The IBM System/370 vector architecture. *IBM Systems Journal*, 25:51–62.

6. Clark, R.S. and T.L. Wilson. 1986. Vector performance of the IBM System 3090. *IBM Systems Journal*, 25(1):63–82.

7. Cray. 1988. Cray X-MP Computer Reference Systems: Hardware Reference Manual. Cray Research Inc., Mendota Heights, Minnesota.

8. CDC. 1987. Cyber 200 Model 205 Computer System: Hardware Reference Manual. Control Data Corporation, St. Paul, Minnesota.

9. Dongara, J.J. and A. Hinds. 1989. Comparison of the Cray X-MP4, Fujitsu VP-2000, and Hitachi S-810/20. In: K. Hwang and D. DeGroot, Eds., *Parallel Processing for Supercomputers and Artificial Intelligence* (McGraw-Hill, New York) pp 289–324.

10. Ercegovac, M.D. and T. Lang. 1986. Vector processing. In: S. Fernbach, Ed., *Supercomputers: Class VI Systems, Hardware and Software* (North-Holland, Amsterdam), pp 29–58.

11. Eoyang, C., R. Menedez, and O.M. Lubeck. 1988. The birth of the second generation: the Hitachi S-820/80. In: *Proceedings, International Symposium on Supercomputing*, pp 296–303.

12. Fong, K. and T.L. Jordan. 1977. Some linear algebraic algorithms and their performance on the Cray-1. Technical report LA-6774, Los Alamos Scientific Laboratory, Los Alamos, New Mexico

13. Gentzsch, W. 1985. *Vectorization of Computer Programs*. Vieweg, Braunschweig, Germany.

14. Hack, J.J. 1986. Peak vs. sustained performance in highly concurrent vector machines. *IEEE Computer*, December:11–19.

15. Harper, D.T. and J.R. Jump. 1987. Vector access performance in parallel memories using a skewed storage scheme. *IEEE Transactions on Computers*, C-36(12):1440–1449.

16. Hockney, R.W. and C.R. Jesshope. 1988. *Parallel Computers 2*. Adam Hilger, Bristol, U.K.

17. Hsu, W.-C. and J.E. Smith. 1993. Performance of DRAM organization in vector supercomputers. In: *Proceedings, 20th International Symposium on Computer Architecture*, pp 327–336.

18. Ibbett, R.N. and N.P. Topham. 1989. *The Architecture of High Performance Computers* (Springer-Verlag, New York), volume 1, Chapters 7–10.

19. Ibbett, R.N., T.M. Hopkins, and K.I.M. McKinnon. 1989. Architectural mechanisms to support sparse vector processing. In: *Proceedings, 16th International Symposium on Computer Architecture*, pp 64–71.

20. Ibbett, R.N., P.C. Capon, and N.P. Topham. 1986. MU6-V: A parallel vector processing system. In: *Proceedings, 12th International Symposium on Computer Architecture*, pp 136–144.

21. Kascic, M.J. 1984. A performance survey of the Cyber 205. In: J.S. Kowalik, Ed., *High-Speed Computation* (Springer-Verlag, New York), pp 191–210.

22. Kuck, D.J. and D.D. Gajksi. 1984. Parallel processing of sparse structures. In: J.S. Kowalik, Ed., *High-Speed Computation* (Springer-Verlag, New York), pp 229–246.

23. Lincoln, N.R. 1986. Technology and design tradeoffs in the creation of a modern supercomputer. In: S. Fernbach, Ed., *Supercomputers: Class VI Systems, Hardware and Software* (North-Holland, Amsterdam), pp 83–112.

24. Lubeck, O., J. Moore, and R. Mendez. 1985. A comparison of three supercomputers: Fujitsu VP-2000, Hitachi S810/200, and Cray X-MP2. *IEEE Computer*, December:10–24.

25. McIntyre, W.. 1987. Amdahl vector processors. In: J.R. Kirkland and J.P. Poore, Eds., *Supercomputers* (Praegen Publishers, New York), pp 85–108.

26. Moore, B., A. Padegs, R. Smith, W. Buchholz. 1987. Concepts of the System/370 vector architecture. In: *Proceedings, 14th International Symposium on Computer Architecture*, pp 282–288.

27. Morris, D. and R.N. Ibbett. 1979. *The MU5 Computer System*. Springer-Verlag, New York.

28. Miura, K. 1986. Fujitsu's supercomputer: Facom vector processor system. In: S. Fernbach, Ed., *Supercomputers: Class VI Systems, Hardware and Software* (North-Holland, Amsterdam), pp 137–152.

29. Neves, K.W. 1984. Vectorization of scientific software. In: S. Fernbach, Ed., *Supercomputers: Class VI Systems, Hardware and Software* (North-Holland, Amsterdam), pp 277–292.

30. Odaka, T., S. Nagashima, and S. Kawabe. 1986. Hitachi supercomputer S-810 array processor system. In: S. Fernbach, Ed., *Supercomputers: Class VI Systems, Hardware and Software* (North-Holland, Amsterdam), pp 113–136.

31. Padegs, A. B.B. Moore, R. Smith, and W. Buchholz. 1988. The IBM System/370 vector architecture: design considerations. *IEEE Transactions on Computers*, 37(5):509–520.

32. Raghavan, R. and J.P. Hayes. 1993. Reducing interference among vector accesses in interleaved memories. *IEEE Transactions on Computers*, 42(4):471-483.

33. Russell, R.M. 1978. The Cray-1 computer system. *Communications of the ACM*, 21(1):63–72.

34. Simmons, S.L. and H.J. Wasserman. 1988. Performance comparison of the Cray-2 and the Cray X-MP/416 supercomputers. In: *Proceedings, ACM International Conference on Supercomputing*.

35. So, K. and V. Zecca. 1988. Cache performance of vector processors. In: *Proceedings, 15th International Symposium on Computer Architecture*, pp 261–268.

36. Temperton, C. 1984. Fast fourier transforms on the Cyber 205. In: J.S. Kowalik, Ed., *High-Speed Computation* (Springer-Verlag, New York), pp 403–416.

37. Thompson, J.R. 1986. The Cray-1, the Cray X-MP, the Cray-2 and beyond: the supersomputers of Cray Research. In: S. Fernbach, Ed., *Supercomputers: Class VI Systems, Hardware and Software* (North-Holland, Amsterdam), pp 69–82.

38. Watanabe, T., H. Katayama, and A. Iwaya. 1986. Introduction of the NEC super-computer NX system. In: S. Fernbach, Ed., *Supercomputers: Class VI Systems, Hardware and Software* (North-Holland, Amsterdam), pp 153–168.

39. Wolfe, M. 1986. Software optimization for supercomputers. In: S. Fernbach, Ed., *Supercomputers: Class VI Systems, Hardware and Software* (North-Holland, Amsterdam), pp pp 221–239.

40. Yang, Q. 1993. Introducing a new cache design into vector computers. *IEEE Transactions on Computers*, 42(12):1411–1424.

Chapter 7

1. AMD. 1997. AMD-K6 MMX Processor. Advanced Micro Devices, Sunnyvale, California.

2. CDC. 1987. Cyber 200 Model 205 Computer System – Hardware Reference Manual. Control Data Corporation, St. Paul, Minnessota.

3. Christie, D. 1996. Developing the AMD-K5 architecture. *IEEE Micro*, 16(2):16–26.

4. Diefendorff, K. and M. Allen, 1992. Organization of the Motorola 88110 super-scalar RISC microprocessor. *IEEE Micro*, 12(4):40–63.

5. Dwyer, H. and H.C. Torng. 1992. An out-of-order superscalar processor with speculative execution and fast, precise interrupts. In: *Proceedings, 25th International Symposium and Workshop on Microarchitecture*, pp 272–281.

6. Gwenap, L. 1996. Digital 21264 sets new standard. *Microprocessor Report*, 10(14).

7. Hennessy, J., N. Jouppi, and J. Gill. 1981. MIPS: A VLSI processor. In: H.T. Kung, B. Sproull, and G. Steele (Eds.), *VLSI Systems and Computations* (Computer Science Press, Rockville, Maryland), pp 347–346.

8. Hwu, W.-M. and Y.N. Patt. 1987. Checkpoint repair for out-of-order execution machines. *IEEE Transactions on Computers*, C-36(12):1496–1514.

9. Iacobovici, S. 1988. A pipelined interface for high floating-point performance with precise exceptions. *IEEE Micro*, 8(3):77–87.

10. Kumar, A. 1997. The HP PA-8000 RISC CPU. *IEEE Micro*, 17(2):27–32.

11. Morris, D. and R.N. Ibbett. 1979. *The MU5 Computer System.* (Springer-Verlag, New York) pp 30–32.

12. Moudgill, M. and S. Vassiliadis. 1996. Precise interrupts. *IEEE Micro*, 16(1):58–67.

13. Okamoto et al 1988. Design considerations for 32-bit microprocessor TX3. In: *Digest of Papers, COMPCON*, pp 25–29.

14. Popescu, V. et al. 1991. The Metaflow architecture. *IEEE Micro*, June: 10–13, 63–73.

15. Smith, J.E. and A.R. Pleszkun. 1988. Implementing precise interrupts in pipelined processors. *IEEE Transactions on Computers*, 37(5):562–573.

16. Sohi, G. 1990. Instruction issue logic for high-performance interruptible, multiple function unit, pipelined computers. *IEEE Transactions on Computers*, 39(3):349–359.

17. Song, S.P., M. Denman, and J. Chang. 1994. The PowerPC 604 microprocessor. *IEEE Micro*, 14(5):8–17.

18. Torng, H.C. and M. Day. 1993. Interrupt handling for out-of-order execution processors. *IEEE Transactions on Computers*, 42(1):122–127.

19. Ullah, N. and M. Holle. 1993. The MC88110 implementation of precise exceptions in a superscalar architecture. *Computer Architecture News*, 21(1):15–25.

20. Wang, C.-J. and F. Emnett. 1993. Implementing precise interrupts in pipelined RISC processors. *IEEE Micro*, 13(4):36–43.

21. Williams, T., N. Patkar, and G. Shen. 1995. SPARC64: A 64-b 64-active-instruction out-of-order-execution MCM processor. *IEEE Journal of Solid-State Circuits*, 30(11):1215–1226.

22. Yeager, K.C. 1996. The MIPS R10000 superscalar microprocessor. *IEEE Micro*, 16(2):28–40.

Index